SHEEP, SWINE, AND POULTRY;

EMBRACING

THE HISTORY AND VARIETIES OF EACH; THE BEST MODES OF BREEDING; THEIR FEEDING AND MANAGEMENT; TOGETHER WITH THE DISEASES TO WHICH THEY ARE RESPECTIVELY SUBJECT, AND THE APPROPRIATE REMEDIES FOR EACH.

BY ROBERT JENNINGS, V. S.,

PROFESSOR OF PATHOLOGY AND OPERATIVE SURGERY IN THE VETERINARY COLLEGE OF PHILADELPHIA; LATE PROFESSOR OF VETERINARY MEDICINE IN THE AGRICULTURAL COLLEGE OF OHIO; SECRETARY OF THE AMERICAN VETERINARY ASSOCIATION OF PHILADELPHIA; AUTHOR OF "THE HORSE AND HIS DISEASES," "CATTLE AND THEIR DISEASES," ETC., ETC.

With Numerous Illustrations.

PHILADELPHIA:

JOHN E. POTTER & COMPANY,

29, 31, 33, AND 35 NORTH TENTH ST.

PREFACE.

ENCOURAGED by the favorable reception of his former works, the author presents in the following pages what is intended by him as a popular compendium relative to Sheep, Swine, and Poultry.

It would not have been a difficult matter to collect material bearing upon each distinct class sufficient for an entire volume of the present size. Indeed, the main trouble experienced has been the selecting of such facts and suggestions only as seemed to him of paramount practical importance. He has not deemed it advisable to cumber his work with items of information which could be of service to particular sections and localities only; but has rather endeavored to present, in a concise, yet comprehensible shape, whatever is essential to be understood concerning the animals in question.

The amateur stock-raiser and the wealthy farmer will, of course, call to their aid all the works, no matter how expensive or voluminous, which are to be found bearing upon the subject in which they are for the time interested. The present volume can scarcely be expected to fill the niche which such might desire to see occupied.

The author's experience as a veterinary surgeon among the great body of our farmers convinces him that what is needed by them in the premises is a treatise, of convenient size, containing the essential features of the treatment and management of each, couched in language free from technicality or rarely scientific expressions, and fortified by the results of actual experience upon the farm.

Such a place the author trusts this work may occupy. He hopes that, while it shall not be entirely destitute of interest for any, it will prove acceptable, in a peculiar degree, to that numerous and thrifty class of citizens to which allusion has already been made.

The importance of such a work cannot be overrated. Take the subject of sheep for example: the steadily growing demand for woollen goods of every description is producing a great and lucrative development of the wool trade. Even light fabrics of wool are now extensively preferred throughout the country to those of cotton. Our imports of wool from England during the past six years have increased at an almost incredible rate, while our productions of the article during the past few years greatly exceed that of the same period in any portion of our history.

Relative to swine, moreover, it may be said that they form so considerable an item of our commerce that a thorough information as to the best mode of raising and caring for them is highly desirable; while our domestic poultry contribute so much, directly and indirectly, to the comfort and partial subsistence of hundreds of thousands, that sensible views touching that division will be of service in almost every household.

To those who are familiar with the author's previous works upon the Horse and Cattle, it is needless to say any thing as to the method adopted by him in discussing the subject of Diseases. To others he would say, that only such diseases are described as are likely to be actually encountered, and such curatives recommended as his own personal experience, or that of others upon whose judgment he relies, has satisfied him are rational and valuable.

The following works, among others, have been consulted: Randall's Sheep Husbandry; Youatt on Sheep; Goodale's Breeding of Domestic Animals; Allen's Domestic Animals; Stephens's Book of the Farm; Youatt on the Hog; Richardson on the Hog; Dixon and Kerr's Ornamental and Domestic Poultry; Bennett's Poultry Book; and Browne's American Poultry Yard.

To those professional brethren who have so courteously furnished him with valuable information, growing out of their own observation and practice, he acknowledges himself especially indebted; and were he certain that they would not take offence, he would be pleased to mention them here by name.

Should the work prove of service to our intelligent American farmers and stock-breeders as a body, the author's end will have been attained.

CONTENTS.

SHEEP AND THEIR DISEASES.

ILLUSTRATIONS.

PAGE

CONTENTS.

SWINE AND THEIR DISEASES.

ILLUSTRATIONS.

CONTENTS.

POULTRY AND THEIR DISEASES.

ILLUSTRATIONS.

A LEICESTER RAM.

HISTORY AND VARIETIES

WITH a single exception—that of the dog—there is no member of the beast family which presents so great a diversity of size, color, form, covering, and general appearance, as characterizes the sheep; and none occupy a wider range of climate, or subsist on a greater variety of food. This animal is found in every latitude between the Equator and the Arctic circle, ranging over barren mountains and through fertile valleys, feeding upon almost every species of edible forage—the cultivated grasses, clovers, cereals, and roots—browsing on aromatic and bitter herbs alike, cropping the leaves and barks from stunted forest shrubs and the pungent, resinous evergreens. In some parts of Norway and Sweden, when other resources fail, he subsists on fish or flesh during the long, rigorous winter, and, if reduced to necessity, even devours his own wool.

In size, he is diminutive or massive; he has many horns, or but two large or small spiral horns, or is polled or hornless. His tail may be broad, or long, or a mere button, discoverable only by the touch. His covering is long and coarse, or short and hairy, or soft and furry, or fine and spiral. His color varies from white or black to every shade of brown, dun, buff, blue, and gray. This wide diversity results from long domestication under almost every conceivable variety of condition.

Among the antediluvians, sheep were used for sacrificial offerings, and their fleeces, in all probability, furnished them with clothing. Since the deluge their flesh has been a favorite food among many nations. Many of the rude, wandering tribes of the East employ them as beasts of burden. The uncivilized—and, to some extent, the refined—inhabitants of Europe use their milk, not only as a beverage, but for making into cheese, butter, and curds—an appropriation of it which is also noticed by Job, Isaiah, and other Old Testament writers, as well as most of the Greek and Roman authors. The ewe's milk scarcely differs in appearance from that of the cow, though it is generally thicker, and yields a pale, yellowish butter, which is always soft and soon becomes rancid. In dairy regions the animal is likewise frequently employed at the tread-mill or horizontal wheel, to pump water, churn milk, or perform other light domestic work.

The calling of the shepherd has, from time immemorial, been conspicuous, and not wanting in dignity and importance. Abel was a keeper of sheep; as were Abraham and his descendants, as well as most of the ancient patriarchs. Job possessed fourteen thousand sheep. Rachel, the favored mother of the Jewish race, "came with her father's sheep, for she kept them." The seven daughters of the priest of Midian "came and drew water for their father's flocks." Moses, the statesman and lawgiver, "learned in all the wisdom of the Egyptians," busied himself in tending "the flocks of Jethro, his father-in-law." David, too, that sweet singer of Israel and its destined monarch—the Jewish hero, poet, and divine—was a keeper of sheep. To shepherds, "abiding in the field, keeping watch over their flocks by night," came the glad tidings of a Saviour's birth. The Hebrew term for sheep signifies, in its etymology, fruitfulness, abundance, plenty—indicative of the blessings which they were destined to confer upon the human family. In the Holy Scriptures, this animal is the chosen symbol of purity and the gentler virtues, the victim of propitiatory sacrifices, and the type of redemption to fallen man.

Among profane writers, Homer and Hesiod, Virgil and Theocritus, introduce them in their pastoral themes; while their heroes and demi-gods—Hercules and Ulysses, Eneas and Numa—carefully perpetuate them in their domains.

In modern times, they have engaged the attention of the most enlightened nations, whose prosperity has been intimately linked with them, wherever wool and its manufactures have been regarded as essential staples. Spain and Portugal, during the two centuries in which they figured as the most enterprising European countries, excelled in the production and manufacture of wool. Flanders, for a time, took precedence of England in the perfection of the arts and the enjoyments of life; and the latter country then sent what little wool she raised to the former to be manufactured. This being soon found highly impolitic, large bounties were offered by England for the importation of artists and machinery; and by a systematic and thorough course of legislation, which looked to the utmost protection and increase of wool and woollens, she gradually carried their production beyond any thing the world had ever seen.

Of the original breed of this invaluable animal, nothing certain is known; four varieties having been deemed by naturalists entitled to that distinction.

These are, 1. The *Musimon*, inhabiting Corsica, Sardinia, and other islands of the Mediterranean, the mountainous parts of Spain and Greece, and some other regions bordering upon that inland sea. These have been frequently domesticated and mixed with the long-cultivated breeds.

2. The *Argali* ranges over the steppes, or inland plains of Central Asia, northward and eastward to the ocean. They are larger and hardier than the Musimon and not so easily tamed.

3. The *Rocky Mountain Sheep*—frequently called the *Big-horn* by our western hunters—is found on the prairies west of the Mississippi, and throughout the wild, mountainous regions extending through California and Oregon to the Pacific. They are larger than the Argali—which in other respects they resemble—and are probably descended from them, since they could easily cross upon the ice at Behring's Straits, from the

north-eastern coast of Asia. Like the Argali, when caught young they are readily tamed; but it is not known that they have ever been bred with the domestic sheep. Before the country was overrun by the white man, they probably inhabited the region bordering on the Mississippi. Father Hennepin—a French Jesuit, who wrote some two hundred years ago—often speaks of meeting with goats in his travels through the territory which is now embraced by Illinois, Wisconsin, and a portion of Minnesota. The wild, clambering propensities of these animals—occupying, as they do, the giddy heights far beyond the reach of the traveller—and their outer coating of hair—supplied underneath, however, with a thick coating of soft wool—give them much the appearance of goats. In summer they are generally found single; but when they descend from their isolated, rocky heights in winter, they are gregarious, marching in flocks under the guidance of leaders.

ROCKY MOUNTAIN SHEEP.

4. The *Bearded Sheep of Africa* inhabit the mountains of Barbary and Egypt. They are covered with a soft, reddish

hair, and have a mane hanging below the neck, and large locks of hair at the ankle.

Many varieties of the domesticated sheep—that is, all the subjugated species—apparently differ less from their wild namesakes than from each other.

The *fat-rumped* and the *broad-tailed sheep* are much more extensively diffused than any other, and occupy nearly all the south-eastern part of Europe, Western and Central Asia, and Northern Africa. They are supposed, from various passages in the Pentateuch in which "the fat and the rump" are spoken of in connection with offerings, to be the varieties which were propagated by the patriarchs and their descendants, the Jewish race. They certainly give indisputable evidence of remote and continued subjugation. Their long, pendent, drowsy ears, and the highly artificial posterior developments, are characteristic of no wild or recently domesticated race.

This breed consists of numerous sub-varieties, differing in all their characteristics of size, fleece, color, etc., with quite as many and marked shades of distinction as the modern European varieties. In Madagascar, they are covered with hair; in the south of Africa, with coarse wool; in the Levant, and along the Mediterranean, the wool is comparatively fine; and from that of the fat-rumped sheep of Thibet the exquisite Cashmere shawls of commerce are manufactured. Both rams and ewes are sometimes bred with horns, and sometimes without, and they exhibit a great diversity of color. Some yield a carcass of scarcely thirty pounds, while others have weighed two hundred pounds dressed. The tail or rump varies greatly, according to the purity and style of breeding; some are less than one-eighth, while others exceed one-third of the entire dressed weight. The fat of the rump or tail is esteemed a great delicacy; in hot climates resembling oil, and in colder, suet.

It is doubtful whether sheep are indigenous to Great Britain; but they are mentioned as existing there at very early periods.

AMERICAN SHEEP.

In North America, there are none, strictly speaking, except the Rocky Mountain breed, already mentioned. The broad-tailed sheep of Asia and Africa were brought into the United States about seventy years ago, under the name of the Tunisian Mountain sheep, and bred with the native flocks. Some of them were subsequently distributed among the farmers of Pennsylvania, and their mixed descendants were highly prized as prolific, and good nurses, coming early to maturity, attaining large weight, of a superior quality of carcass, and yielding a heavy fleece of excellent wool. The principal objection made to them was the difficulty of propagation, which always required the assistance of the shepherd. The lambs were dropped white, red, tawny, bluish, or black; but all, excepting the black, grew white as they approached maturity, retaining some spots of the original color on the cheeks and legs, and sometimes having the entire head tawny or black. The few which descended from the original importations have become blended with American flocks, and have long ceased to be distinguishable from them. The common sheep of Holland were early imported by the Dutch emigrants, who originally colonized New York; but they, in like manner, have long since ceased to exist as a distinct variety.

Improved European breeds have been so largely introduced

during the present century, that the United States at present possesses every known breed which could be of particular benefit to its husbandry. By the census of 1860, there were nearly twenty-three and a half millions of sheep in this country, yielding upwards of sixty and a half million pounds of wool. An almost infinite variety of crosses have taken place between the Spanish, English, and "native" families; carried, indeed, to such an extent that there are, comparatively speaking, few flocks in the United States that preserve entire the distinctive characteristics of any one breed, or that can lay claim to unmixed purity of blood.

The principal breeds in the United States are the so-called "Natives;" the Spanish and Saxon Merinos, introduced from the countries whose names they bear: the New Leicester, or Bakewell; the South-Down; the Cotswold; the Cheviot; and the Lincoln—all from England.

NATIVE SHEEP.

This name is popularly applied to the common coarse-woolled sheep of the country, which existed here previously to the importation of the improved breeds. These were of foreign and mostly of English origin, and could probably claim a common descent from no one stock. The early settlers, emigrating from different sections of the British Empire, and a portion of them from other parts of Europe, brought with them, in all probability, each the favorite breed of his own immediate neighborhood, and the admixture of these formed the mongrel family now under consideration. Amid the perils of war and the incursions of beasts of prey, they were carefully preserved. As early as 1676, New England was spoken of as "abounding with sheep."

These common sheep yielded a wool suitable only for the coarsest fabrics, averaging, in the hands of good farmers, from three to three and a half pounds of wool to the fleece. They were slow in arriving at maturity, compared with the improved English breeds, and yielded, when fully grown, from ten to fourteen pounds of a middling quality of mutton to the quarter. They were usually long-legged, light in the fore-quarter, and narrow on the breast and back; although some rare instances might be found of flocks with the short legs, and some approximation to the general form of the improved breeds. They were excellent breeders, often rearing, almost entirely destitute of care, and without shelter, one hundred per cent. of lambs; and in small flocks, a still larger proportion. These, too, were usually dropped in March, or the earlier part of April. Restless in their disposition, their impatience of restraint almost equalled that of the untamed Argali, from which they were descended; and in many sections of the country it was common to see from twenty to fifty of them roving, with little regard to enclosures, over the possessions of their owner and his neighbors, leaving a large portion of their wool adhering to bushes and thorns, and the remainder placed nearly beyond the possibility of carding, by the tory-weed and burdock, so common on new lands.

To this general character of the native flocks, there was but one exception—a considerably numerous and probably accidental variety, known as the *Otter breed*, or *Creepers*. These were excessively duck-legged, with well-formed bodies, full chests, broad backs, yielding a close, heavy fleece, of medium quality of wool. They were deserved favorites where indifferent stone or wood fences existed, since their power of locomotion was absolutely limited to their enclosures, if protected by a fence not less than two feet high. The quality of their mutton equalled, while their aptitude to fatten was decidedly superior to, their longer-legged contemporaries. The race is now quite extinct.

An excellent variety, called the Arlington sheep, was produced by General Washington, from a cross of a Persian ram upon the Bakewell, which bore wool fourteen inches in length, soft, silky, and admirably suited to combing. These, likewise, have long since become incorporated with the other flocks of the country.

The old common stock of sheep, as a distinct family, have nearly or quite disappeared, owing to universal crossing, to a greater or less extent, with the foreign breeds of later introduction. The first and second cross with the Merino resulted in a decided improvement, and produced a variety exceedingly valuable for the farmer who rears wool solely for domestic purposes. The fleeces are of uneven fineness, being hairy on the thighs, dew-lap, etc.; but the general quality is much improved, the quantity is considerably augmented, the carcass is more compact and nearer the ground, and they have lost their unquiet and roving propensities. The cross with the Saxon, for reasons hereafter to be given, has not generally been so successful. With the Leicester and Downs, the improvement, so far as form size, and a propensity to take on fat are concerned, is manifest.

THE SPANISH MERINO.

The Spanish sheep, in different countries, has, either directly or indirectly, effected a complete revolution in the character of the fleece. The race is unquestionably one of the most ancient extant. The early writers on agriculture and the veterinary art describe various breeds of sheep as existing in Spain, of different colors—black, red, and tawny. The black sheep yielded a fine fleece, the finest of that color which was then known; but the red fleece of Bætica—a considerable part of the Spanish coast on the Mediterranean, comprising the modern Spanish provinces of Gaen, Cordova, Seville, Andalusia, and Granada, which was early colonized by the enterprising Greeks—was, according to Pliny, of still superior quality, and "had no fellow."

A MERINO RAM.

These sheep were probably imported from Italy, and of the Tarentine breed, which had gradually spread from the coast of Syria, and of the Black Sea, and had then reached the western extremity of Europe. Many of them mingled with and improved the native breeds of Spain, while others continued to exist as a distinct race, and, meeting with a climate and an herbage suited to them, retained their original character and value, and were the progenitors of the Merinos of the present day. Columella, a colonist from Italy, and uncle of the writer of an excellent work on agriculture, introduced more of the Tarentine sheep into Bætica, where he resided in the reign of the Emperor Claudius, in the year 41, and otherwise improved on the native breed; for, struck with the beauty of some African rams which had been brought to Rome to be exhibited at the public games, he purchased them, and conveyed them to his farm in Spain, whence, probably, originated the better varieties of the long-woolled breeds of that country.

Before his time, however, Spain possessed a valuable breed; since Strabo, who flourished under Tiberius, speaking of the beautiful woollen cloths that were worn by the Romans, says that the wool was brought from Truditania, in Spain.

The limited region of Italy—overrun, as it repeatedly was, by hordes of barbarians during and after the times of the latest emperors—soon lost her pampered flocks; while the extended regions of Spain—intersected in every direction by almost impassable mountains—could maintain their more hardy race, in defiance of revolution or change.

To what extent the improvements which have been noticed were carried is unknown; but as Spain was at that time highly civilized, and as agriculture was the favorite pursuit of the greater part of the colonists that spread over the vast territory, which then acknowledged the Roman power, it is highly probable that Columella's experiments laid the foundation for a general improvement in the Spanish sheep—an improvement, moreover, which was not lost, nor even materially impaired, during the darker ages that succeeded.

The Merino race possess inbred qualities to an extent surpassed by no others. They have been improved in the general weight and evenness of their fleece, as in the celebrated flock of Rambouillet; in the uniformity and excessive fineness of the fibre, as in the Saxons; and in their form and feeding qualities, in various countries; but there has never yet been deterioration, either in quantity or quality of fleece or carcass, wherever they have been transported, if supplied with suitable food and attention. Most sheep annually shed their wool if unclipped; while the Merino retains its fleece, sometimes for five years, when allowed to remain unshorn.

Conclusive evidence is thus afforded of continued breeding among themselves, by which the very constitution of the wool-producing organs beneath the skin have become permanently established; and this property is transmitted to a great extent, even among the crosses, thus marking the Merino as an ancient and peculiar race.

The remains of the ancient varieties of color, also, as noticed by Pliny, Solinus, and Columella, may still be discovered in the modern Merino. The plain and indeed the only reason that can be assigned for the union of black and gray faces with white bodies, in the same breed, is the frequent intermixture of black and white sheep, until the white prevails in the fleece, and the black is confined to the face and legs. It is still apt to break out occasionally in the individual, unless it is fixed and concentrated in the face and legs, by repeated crosses and a careful selection; and, on the contrary, in the Merino South-Down the black may be reduced by a few crosses to small spots about the legs, while the Merino hue overspreads the countenance. This hue—variously described as a velvet, a buff, a fawn, or a satin-colored countenance, but in which a red tinge not infrequently predominates, still indicates the original colors of the indigenous breeds of Spain; and the black wool, for which Spain was formerly so much distinguished, is still inclined to break out occasionally in the legs and ears of the Merino. In some flocks half the ear is invariably brown, and a coarse black hair is often discernible in the finest pile.

The conquest, in the eighth century, by the Moors of those fine provinces in the south of Spain, so far from checking, served rather to encourage the production of fine wool. The conquerors were not only enterprising, but highly skilled in the useful arts, and carried on extensive manufactories of fine woollen goods, which they exported to different countries. The luxury of the Moorish sovereigns has been the theme of many writers; and in the thirteenth century, when the woollen manufacture flourished in but few places, there were found in Seville no less than sixteen thousand looms. A century later, Barcelona, Perpignan, and Tortosa were celebrated for the fineness of their cloths, which became staple articles of trade throughout the greater part of Europe, as well as on the coast of Africa.

A SPANISH SHEEP DOG.

After the expulsion of the Moors, in the fifteenth century,

by Ferdinand and Isabella, the woollen manufacture languished, and was, in a great degree, lost to Spain, owing to the rigorous banishment of nearly one million industrious Moors, most of whom were weavers. As a consequence, the sixteen thousand looms of Seville dwindled down to sixty. The Spanish government perceived its fatal mistake too late, and subsequent efforts to gain its lost vantage-ground in respect to this manufacture proved fruitless. During all that time, however, the Spanish sheep appear to have withstood the baneful influence of almost total neglect; and although the Merino flocks and Merino wool have improved under the more careful management of other countries, the world is originally indebted to Spain for the most valuable material in the manufacture of cloth.

The perpetuation of the Merino sheep in all its purity, amid the convulsions which changed the entire political framework of Spain and destroyed every other national improvement, strikingly illustrates the primary determining power of blood or breeding, as well as the agency of soil and climate—possibly too much underrated in modern times.

These Spanish sheep are divided into two classes: the *stationary*, or those that remain during the whole of the year on a certain farm, or in a certain district, there being a sufficient provision for them in winter and in summer; and the *migratory*, or those which wander some hundreds of miles twice in the year, in quest of pasturage. The principal breed of stationary sheep consists of true Merinos; but the breeds most sought for, and with which so many countries have been enriched, are the Merinos of the migratory description, which pass the summer in the mountains of the north, and the winter on the plains toward the south of Spain.

The first impression made by the Merino sheep on one unacquainted with its value would be unfavorable. The wool lying closer and thicker over the body than in most other breeds, and being abundant in yolk—or a peculiar secretion from the glands of the skin, which nourishes the wool and causes it to mat closely together—is covered with a dirty crust, often full of cracks. The legs are long, yet small in the bone; the breast and the back are narrow, and the sides somewhat flat; the fore-shoulders and bosoms are heavy, and too much of their weight is carried on the coarser parts. The horns of the male are comparatively large, curved, and with more or less of a spiral form; the head is large, but the forehead rather low. A few of the females are horned; but, generally speaking, they are without horns. Both male and female have a peculiar coarse and unsightly growth of hair on the forehead and cheeks, which the careful shepherd cuts away before the shearing-time; the other part of the face has a pleasing and characteristic velvet appearance. Under the throat there is a singular looseness of skin, which gives them a remarkable appearance of throatiness, or hollowness in the neck. The pile or hair, when pressed upon, is hard and unyielding, owing to the thickness into which it grows on the pelf, and the abundance of the yolk, retaining all the dirt and gravel which falls upon it; but, upon examination, the fibre exceeds, in fineness and in the number of serrations and curves, that which any other sheep in the world produces. The average weight of the fleece in Spain is eight pounds from the ram, and five from the ewe. The staple differs in length in different provinces. When fatted, these sheep will weigh from twelve to sixteen pounds per quarter.

The excellence of the Merinos consist in the unexampled fineness and felting property of their wool, and in the weight of it yielded by each individual sheep; the closeness of that wool, and the luxuriance of the yolk, which enable them to support extremes of cold and wet quite as well as any other breed; the readiness with which they adapt themselves to every change of climate, retaining, with common care, all their fineness of wool, and thriving under a burning tropical sun, and in the frozen regions of the north; an appetite which renders them apparently satisfied with the coarsest food; a quietness and patience into whatever pasture they are turned; and a gentleness and tractableness not excelled in any other breed.

Their defects—partly attributable to the breed, but more to the improper mode of treatment to which they are occasionally subjected—are, their unthrifty and unprofitable form; a tendency to abortion, or barrenness; a difficulty of yeaning, or giving birth to their young; a paucity of milk; and a too frequent neglect of their lambs. They are likewise said, notwithstanding the fineness of their wool, and the beautiful red color of the skin when the fleece is parted, to be more subject to cutaneous affections than most other breeds. Man, however, is far more responsible for this than Nature. Every thing was sacrificed in Spain to fineness and quantity of wool. These were supposed to be connected with equality of temperature, or, at least, with freedom from exposure to cold; and, therefore, twice in the year, a journey of four hundred miles was undertaken, at the rate of eighty or a hundred miles per week—the spring journey commencing when the lambs were scarcely four months old. It is difficult to say in what way the wool of the migratory sheep was, or could be, benefited by these periodical journeys. Although among them is found the finest and most valuable wool in Spain, yet the stationary sheep, in certain provinces—Segovia, Leon, and Estremadura—are more valuable than the migratory flocks of others. Moreover, the fleece of some of the German Merinos—which do not travel at all, and are housed all the winter—greatly exceeds that obtained from the best migratory breed—the Leonese—in fineness and felting property; and the wool of the migratory sheep has been, comparatively speaking, driven out of the market by that from sheep which never travel. With respect to the carcass, these harassing journeys, occupying one-quarter of the year, tend to destroy all possibility of fattening, or any tendency toward it, and the form and the constitution of the flock are deteriorated, and the lives of many sacrificed.

The first importation of Merinos into the United States took place in 1801; a banker of Paris, Mr. Delessert, having shipped four, of which but one arrived in safety at his farm near Kingston, in New York; the others perished on the passage. The same year, Mr. Seth Adams, of Massachusetts, imported a pair from France. In 1802, Chancellor Livingston, then American Minister at the court of Versailles, sent two choice pairs from the Rambouillet flock—which was started, in 1786, by placing four hundred ewes and rams, selected from the choicest Spanish flocks, on the royal farm of that name, in France—to Claremont, his country-seat, on the Hudson river. In the latter part of the same year, Colonel Humphreys, American Minister to Spain, shipped two hundred, on his departure from that country. The largest importations, however, were made through Hon. William Jarvis, of Vermont, then American Consul at Lisbon, Portugal, in 1809, 1810, and 1811, who succeeded in obtaining the choicest sheep of that country. Various subsequent importations took place, which need not be particularized.

The cessation of all commercial intercourse with England, in 1808 and 1809, growing out of difficulties with that country, directed attention, in an especial manner, toward manufacturing and wool-growing. The Merino, consequently, rose into importance, and so great was the interest aroused, that from a thousand to fourteen hundred dollars a head was paid for them. Some of the later importations, unfortunately, arrived in the worst condition, bringing with them those scourges of the sheep family, the scab and the foot-rot; which evils, together with increased supply, soon brought them down to less than a twentieth part of their former price. When, however, it was established, by actual experiment, that their wool did not deteriorate in this country, as had been feared by many, and that they became readily acclimated, they again rose into favor. The prostration of the manufacturing interests of the country, which ensued soon afterwards, rendered the Merino of comparatively little value, and ruined many who had purchased them at their previous high prices. Since that period, the valuation of the sheep which bear the particular wool has, as a matter of course, kept pace with the fluctuations in the price of the wool.

The term Merino, it must be remembered, is but the general appellation of a breed, comprising several varieties, presenting essential points of difference in size, form, quality and quantity of wool. These families have generally been merged, by interbreeding, in the United States and other countries which have received the race from Spain. Purity of *Merino* blood, and actual excellence in the individual and its ancestors, form the only standard in selecting sheep of this breed. Families have, indeed, sprung up in this country, exhibiting wider points of difference than did those of Spain. This is owing, in some cases, doubtless, to particular causes of breeding; but more often, probably, to concealed or forgotten infusions of other blood. The question, which has been at times raised, whether there are any Merinos in the United States, descendants of the early importations, of unquestionable purity of blood, has been conclusively settled in the affirmative.

The minor distinctions among the various families into which, as has already been intimated, the American Merino has diverged, are numerous, but may all, perhaps, be classed under three general heads.

The *first* is a large, short-legged, strong, exceedingly hardy sheep, carrying a heavy fleece, ranging from medium to fine, free from hair in properly bred flocks; somewhat inclined to throatiness, but not so much so as the Rambouillets; bred to exhibit external concrete gum in some flocks, but not commonly so; their wool rather long on back and belly, and exceedingly dense; wool whiter within than the Rambouillets; skin the same rich rose-color. Sheep of this class are larger and stronger than those originally imported, carry much heavier fleeces, and in well-selected flocks, or individuals, the fleece is of a decidedly better quality.

The *second* class embraces smaller animals than the preceding; less hardy; wool, as a general thing, finer, and covered with a black, pitchy gum on its extremities; fleece about one-fourth lighter than in the former class.

The *third* class, bred at the South, mostly, includes animals still smaller and less hardy, and carrying still finer and lighter fleeces. The fleece is destitute of external gum. The sheep and wool have a close resemblance to the Saxon; and, if not actually mixed with that blood, they have been formed into a similar variety, by a similar course of breeding.

The mutton of the Merino, notwithstanding the prejudices existing on the subject, is short-grained, and of good flavor, when killed at a proper age, and weighs from ten to fourteen pounds to the quarter. It is remarkable for its longevity, retaining its teeth, and continuing to breed two or three years longer than the common sheep, and at least half a dozen years longer than the improved English breeds. It should, however, be remarked, in this connection, that it is correspondingly slow in arriving at maturity, as it does not attain its full growth before three years of age; and the ewes, in the best managed flocks, are rarely permitted to breed before they reach that age.

OUT AT PASTURE.

The Merino is a far better breeder than any other fine-woolled sheep, and its lambs, when newly dropped, are claimed to be hardier than the Bakewell, and equally so with the high-bred South-Down. The ewe, as has been intimated, is not so good a nurse, and will not usually do full justice to more than one lamb. Eighty or ninety per cent. is about the ordinary number of lambs reared, though it often reaches one hundred per cent., in carefully managed or small flocks.

Allusion has heretofore been made to the cross between the Merino and the native sheep. On the introduction of the Saxon family of the Merinos, they were universally engrafted on the parent stock, and the cross was continued until the Spanish blood was nearly bred out. When the admixture took place with judiciously selected Saxons, the results were not unfavorable for certain purposes. These instances of judicious crossing were, unfortunately, rare. Fineness of wool was made the only tests of excellence, no matter how scanty its quantity, or how diminutive or miserable the carcass. The consequence was, as might be supposed, the ruin of most of the Merino flocks.

THE SAXON MERINO.

The indigenous breed of sheep in Saxony resembled that of the neighboring states, and consisted of two distinct varieties—one bearing a wool of some value, and the other yielding a fleece applicable only to the coarsest manufactures.

At the close of the seven years war, Augustus Frederic, the Elector of Saxony, imported one hundred rams and two hundred ewes from the most improved Spanish flocks, and placed a part of them on one of his own farms, in the neighborhood of Dresden, which he kept unmixed, as he desired to ascertain how far the pure Spanish breed could be naturalized in that country The other part of the flock was distributed on other farms, and devoted to the improvement of the Saxon sheep.

It was soon sufficiently apparent that the Merinos did not degenerate in Saxony. Many parcels of their wool were not inferior to the choicest Leonese fleeces. The best breed of the native Saxons was also materially improved. The majority of the shepherds were, however, obstinately prejudiced against the innovation; but the elector, resolutely bent upon accomplishing his object, imported an additional number, and compelled the crown-tenants, then occupying lands under him, to purchase a certain number of the sheep.

Compulsion was not long necessary; the true interest of the shepherds was discovered; pure Merinos rapidly increased in Saxony, and became perfectly naturalized. Indeed, after a considerable lapse of years, the fleece of the Saxon sheep began, not only to equal the Spanish, but to exceed it in fineness and manufacturing value. To this result the government very materially contributed, by the establishment of an agricultural school, and other minor schools for shepherds, and by distributing various publications, which plainly and intelligibly showed the value and proper management of the Merino. The breeders were selected with almost exclusive reference to the quality of the fleece. Great care was taken to prevent exposure throughout the year, and they were housed on every slight emergency. By this course of breeding and treatment the size and weight of the fleece were reduced, and that hardiness and vigor of constitution, which had universally characterized the migratory Spanish breed, were partially impaired. In numerous instances, this management resulted in permanent injury to the character of the flocks.

The first importation of Saxons into this country was made in 1823, by Samuel Heustan, a merchant of Boston, Massachusetts, and consisted of four good rams, of which two went to Boston, and the others to Philadelphia. The following year, seventy-seven—about two-thirds of which number only were pure-blooded—were brought to Boston, sold at public auction at Brooklyn, N.Y., as "pure-blooded electoral Saxons," and thus scattered over the country. Another lot, composed of grade sheep and pure-bloods, was disposed of, not long afterwards, by public sale, at Brighton, near Boston, and brought increased prices, some of them realizing from four hundred to five hundred and fifty dollars.

These prices gave rise to speculation, and many animals, of a decidedly inferior grade, were imported, which were thrown upon the market for the most they could command. The sales in many instances not half covering the cost of importation, the speculation was soon abandoned. In 1827, Henry D. Grove, of Hoosic, N. Y., a native of Germany, and a highly intelligent and thoroughly bred shepherd, who had accompanied some of the early importations, imported one hundred and fifteen choice animals for his own breeding, and, in the following year, eighty more. These formed the flock from which Mr. Grove bred, to the time of his decease, in 1844. The average weight of fleece from his entire flock, nearly all of which were ewes and lambs, was ten pounds and fourteen ounces, thoroughly washed on the sheep's back. This was realized after a short summer and winter's keep, when the quantity of hay or its equivalent fed to the sheep did not exceed one and a half pounds, by actual weight, per day, except to the ewes, which received an additional quantity just before and after lambing. This treatment was attended with no disease or loss by death, and with an increase of lambs, equalling one for every ewe.

The Saxon Merino differs materially in frame from the Spanish; there is more roundness of carcass and fineness of bone, together with a general form and appearance indicative of a disposition to fatten. Two distinct breeds are noticed. One variety has stouter legs, stouter bodies, head and neck comparatively short and broad, and body round; the wool grows most on the face and legs; the grease in the wool is almost pitchy. The other breed, called Escurial, has longer legs, with a long, spare neck and head; very little wool on the latter; and a finer, shorter, and softer character in its fleece, but less in quantity.

From what has just been stated it will be seen that there are few Saxon flocks in the United States that have not been reduced to the quality of grade sheep, by the promiscuous admixture of the pure and the impure which were imported together; all of them being sold to our breeders as pure stock. Besides, there are very few flocks which have not been again crossed with the Native or the Merino sheep of our country, or with both. Those who early purchased the Merino crossed them with the Native; and when the Saxons arrived those mongrels were bred to Saxon rams. This is the history of three-quarters, probably, of the Saxon flocks of the United States.

As these sheep have now so long been bred toward the

Saxon that their wool equals that of the pure-bloods, it may well be questioned whether they are any worse for the admixture; when crossed only with the Merino, it is, undoubtedly, to their advantage. The American Saxon, with these early crosses in its pedigree, is, by general admission, a hardier and more easily kept animal than the pure Escurial or Electoral Saxon. Climate, feed, and other causes have, doubtless, conspired, as in the case of the Merino, to add to their size and vigor; but, after every necessary allowance has been made, they generally owe these qualities to those early crosses.

The fleeces of the American Saxons weigh, on the average, from two or two and a quarter to three pounds. They are, comparatively speaking, a tender sheep, requiring regular supplies of good food, good shelter in winter, and protection in cool weather from storms of all kinds; but they are evidently hardier than the parent German stock. In docility and patience under confinement, in late maturity and longevity, they resemble the Merinos, from which they are descended; though they do not mature so early as the Merino, nor do they ordinarily live so long. They are poorer nurses; their lambs are smaller, fatter, and far more likely to perish, unless sheltered and carefully watched; they do not fatten so well, and, being considerably lighter, they consume an amount of food considerably less.

Taken together, the American Saxons bear a much finer wool than the American Merinos; though this is not always the case, and many breeders of Saxons cross with the Merino, for the purpose of increasing the weight of their fleeces without deteriorating its quality. Our Saxon wool, as a whole, falls considerably below that of Germany; though individual specimens from Saxons in Connecticut and Ohio compare well with the highest German grades. This inferiority is not attributable to climate or other natural causes, or to a want of skill on the part of our breeders; but to the fact that but a very few of our manufacturers have ever felt willing to make that discrimination in prices which would render it profitable to breed those small and delicate animals which produce this exquisite quality of wool.

THE NEW LEICESTER.

The unimproved Leicester was a large, heavy, coarse-woolled breed of sheep, inhabiting the midland counties of England. It was a slow feeder, its flesh coarse-grained, and with little flavor. The breeders of that period regarded only size and weight of fleece.

A COUNTRY SCENE.

About the middle of the last century, Robert Bakewell, of Dishley, in Leicestershire, first applied himself to the improvement of the sheep in that country. Before his improvements, aptitude to fatten and symmetry of shape—that is, such shape as should increase as much as possible the most valuable parts of the animal, and diminish the offal in the same proportion—were entirely disregarded. Perceiving that smaller animals increased in weight more rapidly than the very large ones, that they consumed less food, that the same quantity of herbage, applied to feeding a large number of small sheep, would produce more meat than when applied to feeding the smaller number of large sheep, which alone it would support, and that sheep carrying a heavy fleece of wool possessed less propensity to fatten than those which carried one of a more moderate weight, he selected from the different flocks in his neighborhood, without regard to size, the sheep which appeared to him to have the greatest propensity to fatten, and whose shape possessed the peculiarities which, in his judgment, would produce the largest proportion of valuable meat, and the smallest quantity of bone and offal.

He was also of opinion that the first object to be attended to in breeding sheep is the value of the carcass, and that the fleece ought always to be a secondary consideration; and this for the obvious reason that, while the addition of two or three pounds of wool to the weight of a sheep's fleece is a difference of great amount, yet if this increase is obtained at the expense of the animal's propensity to fatten, the farmer may lose by it ten or twelve pounds of mutton.

The sort of sheep, therefore, which he selected were those possessed of the most perfect symmetry, with the greatest aptitude to fatten, and rather smaller in size than the sheep generally bred at that time. Having formed his stock from sheep so selected, he carefully attended to the peculiarities of the individuals from which he bred, and, so far as can be ascertained—for all of Mr. Bakewell's measures were kept secret, even from his most intimate friends, and he died without throwing, voluntarily, the least light on the subject—did not object to breeding from near relations, when, by so doing, he brought together animals likely to produce a progeny possessing the characteristics which he wished to obtain.

Having thus established his flock, he adopted the practice—which has since been constantly followed by the most eminent breeders of sheep—of letting rams for the season, instead of selling them to those who wished for their use. By this means the ram-breeder is enabled to keep a much larger number of rams in his possession; and, consequently, his power of selecting those most suitable to his flock, or which may be required to correct any faults in shape or quality which may occur in it, is greatly increased. By cautiously using a ram for one season, or by observing the produce of a ram let to some other breeder, he can ascertain the probable qualities of the lambs which such ram will get, and thus avoid the danger of making mistakes which would deteriorate the value of his stock. The farmers, likewise, who hire the rams, have an opportunity of varying the rams from which they breed much more than they otherwise could do; and they are also enabled to select from sheep of the best quality, and from those best calculated to effect the greatest improvement in their flocks.

The idea, when first introduced by him, was so novel that he had great difficulty in inducing the farmers to act upon it; and his first ram was let for sixteen shillings. So eminent, however, was his success, that, in 1787, he let three rams, for a single season, for twelve hundred and fifty pounds (about six thousand two hundred dollars), and was offered ten hundred and fifty pounds (about five thousand two hundred dollars) for twenty ewes. Soon afterwards he received the enormous price of eight hundred guineas (or four thousand dollars) for two-thirds of the services of a ram for a single season, reserving the other third for himself.

The improved Leicester is of large size, but somewhat smaller than the original stock, and in this respect falls considerably below the coarser varieties of Cotswold, Lincoln, etc. Where there is a sufficiency of feed, the New Leicester is unrivalled for its fattening propensities; but it will not bear hard stocking, nor must it be compelled to travel far in search of its food. It is, in fact, properly and exclusively a lowland sheep. In its appropriate situation—on the luxuriant herbage of the highly cultivated lands of England—it possesses unequalled earliness of maturity; and its mutton, when not too fat, is of a good quality, but is usually coarse, and comparatively deficient in flavor, owing to that unnatural state of fatness which it so readily assumes, and which the breeder, to gain weight, so generally feeds for. The wethers, having reached their second year, are turned off in the succeeding February or March, and weigh at that age from thirty to thirty-five pounds to the quarter. The wool of the New Leicester is long, averaging, after the first shearing, about six inches; and the fleece of the American animal weighs about six pounds. It is of coarse quality, and little used in the manufacture of cloth, on account of its length, and that deficiency of felting properties common, in a greater or less extent, to all English breeds. As a combing wool, however, it stands first, and is used in the manufacture of the finest worsteds, and the like textures.

The high-bred Leicesters of Mr. Bakewell's stock became

shy breeders and poor nurses; but crosses subsequently adopted have, to some extent, obviated these defects. The lambs are not, however, generally regarded as very hardy, and they require considerable attention at the time of yeaning, particularly if the weather is even moderately cold or stormy. The grown sheep, too, are much affected by sudden changes in the weather; an abrupt change to cold being pretty certain to be registered on their noses by unmistakable indications of catarrh or "snuffles."

In England, where mutton is generally eaten by the laboring classes, the meat of this variety is in very great demand; and the consequent return which a sheep possessing such fine feeding qualities is enabled to make renders it a general favorite with the breeder. Instances are recorded of the most extraordinary prices having been paid for these animals. They have spread into all parts of the British dominions, and been imported into the other countries of Europe and into the United States.

They were first introduced into our own country, some forty years since, by Christopher Dunn, of Albany, N. Y. Subsequent importations have been made by Mr. Powel, of Philadelphia, and various other gentlemen. The breed, however, has never proved a favorite with any large class of American farmers. Our long, cold winters—but, more especially, our dry, scorching summers, when it is often difficult to obtain the rich, green, tender feed in which the Leicester delights—together with the general deprivation of green feed in the winter, rob it of its early maturity, and even of the ultimate size which it attains in England. Its mutton is too fat, and the fat and lean are too little intermixed to suit

American taste. Its wool is not very salable, owing to the dearth of worsted manufactures in our country. Its early decay and loss of wool constitute an objection to it, in a country where it is often so difficult to advantageously turn off sheep, particularly ewes. But, notwithstanding all these disadvantages, on rich lowland farms, in the vicinity of considerable markets, it will always in all probability make a profitable return.

The head of the New Leicester should be hornless, long, small, tapering towards the muzzle, and projecting horizontally forward; the eyes prominent, but with a quiet expression; the ears thin, rather long, and directed backward; the neck full and broad at its base, where it proceeds from the chest, so that there is, with the slightest possible deviation, one continued horizontal line from the rump to the poll; the breast broad and full; the shoulders also broad and round, and no uneven or angular formation where the shoulders join either the neck or the back—particularly no rising of the withers, or hollow behind the situation of these bones; the arm fleshy throughout its whole extent, and even down to the knee; the bones of the leg small, standing wide apart; no looseness of skin about them, and comparatively void of wool; the chest and barrel at once deep and round; the ribs forming a considerable arch from the spine, so as, in some cases—and especially when the animal is in good condition—to make the apparent width of the chest even greater than the depth; the barrel ribbed well home; no irregularity of line on the back or belly, but on the sides; the carcass very gradually diminishing in width towards the rump; the quarters long and full, and, as with the fore-legs, the muscles extending down to the hock;

the thighs also wide and full; the legs of a moderate length; and the pelt also moderately thin, but soft and elastic, and covered with a good quantity of white wool, not so long as in some breeds, but considerably finer.

THE SOUTH-DOWN.

A long range of chalky hills, diverging from the chalky stratum which intersects England from Norfolk to Dorchester, is termed the South-Downs. They enter the county of Sussex on the west side, and are continued almost in a direct line, as far as East Bourne, where they reach the sea. They may be regarded as occupying a space of more than sixty miles in length, and about five or six in breadth, consisting of a succession of open downs, with few enclosures, and distinguished by their situation and name from a more northern tract of similar elevation and soil, passing through Surrey and Kent, and terminating in the cliffs of Dover, and of the Forelands. On these downs a certain breed of sheep has been produced for many centuries, in greater perfection than elsewhere; and hence have sprung those successive colonies which have found their way abroad and materially benefited the breed of short-woolled sheep wherever they have gone.

A SOUTH-DOWN RAM.

It is only, however, within a comparatively recent period that they have been brought to their present perfection. As recently as 1776 they were small in size, and of a form not superior to the common woolled sheep of the United States; they were far from possessing a good shape, being long and thin in the neck, high on the shoulders, low behind, high on the loins, down on the rump, the tail set on very low, perpendicular from the hip-bones, sharp on the back; the ribs flat, not bowing, narrow in the fore-quarters, but good in the leg, although having big bones. Since that period a course of judicious breeding, pursued by Mr. John Ellman, of Glynde, in Sussex, has mainly contributed to raise this variety to its present value; and that, too, without the admixture of the slightest degree of foreign blood.

This pure, improved family, it will be borne in mind, is spoken of in the present connection; inasmuch as the original stock, presenting, with trifling modifications, the same characteristics which they exhibited seventy-five years ago, are yet to be found in England; and the intermediate space between these two classes is occupied by a variety of grades, rising or falling in value, as they approximate to or recede from the improved blood.

The South-Down sheep are polled, but it is probable that the original breed was horned, as it is not unusual to find among the male South-Down lambs some with small horns. The dusky, or at times, black hue of the head and legs fully establishes the original color of the sheep, and, perhaps of all sheep; while the later period at which it was seriously attempted to get rid of this dingy hue proving unsuccessful, only confirms this view. Many of the lambs have been dropped entirely black.

It is an upland sheep, of medium size, and its wool—which in point of length belongs to the middle class, and differs essentially from Merino wool of any grade, though the fibre in some of the finest fleeces may be of the same apparent fineness with half or one-quarter blood Merino—is deficient in felting properties, making a fuzzy, hairy cloth, and is no longer used in England, unless largely mixed with foreign wool, even for the lowest class of cloths. As it has deteriorated, however, it has increased in length of staple, in that country, to such an extent that improved machinery enables it to be used as a combing-wool, for the manufacture of worsteds. Where this has taken place it is quite as profitable as when it was finer and shorter. In the United States, where the demand for combing-wool is so small that it is easily met by a better article, the same result would not probably follow. Indeed, it may well be doubted whether the proper combing length will be easily reached, or at least maintained in this country, in the absence of that high feeding system which has undoubtedly given the wool its increased length in England. The average weight of fleece in the hill-fed sheep is three pounds; on rich lowlands, a little more.

The South-Down, however, is cultivated more particularly for its mutton, which for quality takes precedence of all other—from sheep of good size—in the English markets. Its early maturity and extreme aptitude to lay on flesh, render it peculiarly valuable for this purpose. It is turned off at the age of two years, and its weight at that age is, in England, from eighty to one hundred pounds. High-fed wethers have reached from thirty-two to even forty pounds a quarter. Notwithstanding its weight, it has a patience of occasional short keep, and an endurance of hard stocking, equal to any other sheep. This gives it a decided advantage over the bulkier Leicesters and Lincolns, as a mutton sheep, in hilly districts and those producing short and scanty herbage. It is hardy and healthy, though, in common with the other English varieties, much subject to catarrh, and no sheep better withstands our American winters. The ewes are prolific breeders and good nurses.

The Down is quiet and docile in its habits, and, though an industrious feeder, exhibits but little disposition to rove. Like the Leicester, it is comparatively a short-lived animal, and the fleece continues to decrease in weight after it reaches maturity. It crosses better with short and middle-woolled breeds than the Leicester. A sheep possessing such qualities, must, of necessity, be valuable in upland districts in the vicinity of markets. The Emperor of Russia paid Mr. Ellman three hundred guineas (fifteen hundred dollars) for two rams; and, in 1800, a ram belonging to the Duke of Bedford was let for one season at eighty guineas (four hundred dollars), two others at forty guineas (two hundred dollars) each, and four more at twenty-eight guineas (one hundred and forty dollars) each. The first importation into the United States was made by Col. J. H. Powell, of Philadelphia. A subsequent importation, in 1834, cost sixty dollars a head.

The desirable characteristics of the South-Down may be thus summed up: The head small and hornless; the face speckled or gray, and neither too long nor too short; the lips thin, and the space between the nose and the eyes narrow; the under-jaw or chap fine and thin; the ears tolerably wide and well-covered with wool, and the forehead also, and the whole space between the ears well protected by it, as a defence against the fly; the eye full and bright, but not prominent; the orbits of the eye, the eye-cap or bone not too projecting, that it may not form a fatal obstacle in lambing; the neck of a medium length, thin toward the head, but enlarging toward the shoulders, where it should be broad and high and straight in its whole course above and below.

The breast should be wide, deep, and projecting forward between the fore-legs, indicating a good constitution and a disposition to thrive; corresponding with this, the shoulders should be on a level with the back, and not too wide above; they should bow outward from the top to the breast, indicating a springing rib beneath, and leaving room for it; the ribs coming out horizontally from the spine, and extending far backward, and the last rib projecting more than others; the back flat from the shoulders to the setting on of the tail; the loin broad and flat; the rump broad, and the tail set on high, and nearly on a level with the spine.

The hips should be wide; the space between them and the last rib on each side as narrow as possible, and the ribs generally presenting a circular form like a barrel; the belly as straight as the back; the legs neither too long nor too short; the fore-legs straight from the breast to the foot, not bending inward at the knee, and standing far apart, both before and behind; the hock having a direction rather outward, and the twist, or the meeting of the thighs behind, being particularly full; the bones fine, yet having no appearance of weakness, and of a speckled or dark color; the belly well defended with wool, and the wool coming down before and behind to the knee and to the hock; the wool short, close, curled and fine, and free from spiry projecting fibres.

THE COTSWOLD.

The Cotswolds, until improved by modern crosses, were a very large, coarse, long-legged, flat-ribbed variety, light in the fore-quarter, and shearing a long, heavy, coarse fleece of wool. They were formerly bred only on the hills, and fatted in the valleys, of the Severn and the Thames; but with the enclosures of the Cotswold hills, and the improvement of their cultivation, they have been reared and fatted in the same district. They were hardy, prolific breeders, and capital nurses; deficient in early maturity, and not possessing feeding properties equalling those of the South-Down or New Leicester.

THE COTSWOLD.

They have been extensively crossed with the Leicester sheep—producing thus the modern or improved Cotswold—by which their size and fleece have been somewhat diminished, but their carcasses have been materially improved, and their maturity rendered earlier. The wethers are sometimes fattened at fourteen months old, when they weigh from fifteen to twenty-four pounds to a quarter; and at two vears old, increase to twenty or thirty pounds.

The wool is strong, mellow, and of good color, though rather coarse. six to eight inches in length, and from seven to eight pounds per fleece. The superior hardihood of the improved Cotswold over the Leicester, and their adaptation to common treatment, together with the prolific nature of the ewes, and their abundance of milk, have rendered them in many places rivals of the New Leicester, and have obtained for them, of late years, more attention to their selection and general treatment, under which management still farther improvement has been made. They have also been used in crossing other breeds, and have been mixed with the Hampshire Downs. Indeed, the improved Cotswold, under the name of new, or improved Oxfordshire sheep, have frequently been the successful candidates for prizes offered for the best long-woolled sheep at some of the principal agricultural meetings or shows in England. The quality of their mutton is considered superior to that of the Leicester; the tallow being less abundant, with a larger development of muscle or flesh.

The degree to which the cross between the Cotswold and Leicester may be carried, must depend upon the nature of the old stock, and on the situation and character of the farm. In exposed situations, and somewhat scanty pasture, the old blood should decidedly prevail. On a more sheltered soil, and on land that will bear closer stocking, a greater use may be made of the Leicester. Another circumstance that should guide the farmer is the object which he has principally in view. If he expects to derive his chief profits from the wool, he will look to the primitive Cotswolds; if he expects to gain more as a grazier, he will use the Leicester ram more freely.

Sheep of this breed, now of established reputation, have been imported into the United States by Messrs. Corning and Gotham, of Albany, and bred by the latter.

THE CHEVIOT.

On the steep, storm-lashed Cheviot hills, in the extreme north of England, this breed first attracted notice for their great hardiness in resisting cold, and for feeding on coarse, heathery herbage. A cross with the Leicester, pretty generally resorted to, constitutes the improved variety.

A CHEVIOT EWE.

The Cheviot readily amalgamates with the Leicester—the rams employed in the system of breeding, which has been extensively introduced for producing the first cross of this descent, being of the pure Leicester breed—and the progeny is superior in size, weight of wool, and tendency to fatten, to the native Cheviot. The benefit, however, may be said to end with the first cross; and the progeny of this mixed descent is greatly inferior to the pure Leicester in form and fattening properties, and to the pure Cheviot in hardiness of constitution.

The improved Cheviot has greatly extended itself throughout the mountains of Scotland, and in many instances supplanted the black-faced breed; but the change, though often advantageous, has in some cases been otherwise—the latter being somewhat hardier, and more capable of subsisting on heathy pasturage. They are a hardy race, however; well suited for their native pastures, bearing, with comparative impunity, the storms of winter, and thriving well on poor keep. The purest specimens are to be found on the Scotch side of the Cheviot hills, and on the high and stony mountain farms which lie between that range and the sources of the Teviot. These sheep are a capital mountain stock, provided the pasture resembles those hills, in containing a good proportion of rich herbage. Though less hardy than the black-faced sheep of Scotland, they are more profitable as respects their feeding, making more flesh on an equal quantity of food, and making it more quickly.

They have white faces and legs, open countenances, lively eyes, and are without horns; the ears are large, and somewhat singular, and there is much space between the ears and eyes; the carcass is long; the back straight; the shoulders rather light; the ribs circular; and the quarters good. The legs are small in the bone, and covered with wool, as well as the body, with the exception of the face. The wether is fit for the butcher at three years old, and averages from twelve to eighteen pounds a quarter; the mutton being of a good quality, though inferior to the South-Down, and of less flavor than the black-faced. The Cheviot, though a mountain breed, is quiet and docile, and easily managed.

The wool is about the quality of Leicester, coarse and long, suitable only for the manufacture of low coatings and flushings. It closely covers the body, assisting much in preserving it from the effects of wet and cold. The fleece averages about three and a half pounds. Formerly, the wool was extensively employed in making cloths; but having given place to the finer Saxony wools, it has sunk in price, and been confined to combing purposes. It has thus become altogether a secondary consideration.

The Cheviots have become an American sheep by their repeated importations into this country. The wool on several choice sheep, imported by Mr. Carmichael, of New York, was from five to seven inches long, coarse, but well suited to combing.

THE LINCOLN.

The old breed of Lincolnshire sheep was hornless, had white faces, and long, thin, and weak carcasses; the ewes weighed from fourteen to twenty pounds a quarter; the three-year old wethers from twenty to thirty pounds; legs thick, rough and white; pelts thick; wool long—from ten to eighteen inches—and covering a slow-feeding, coarse-grained carcass of mutton.

A judicious system of breeding, which avoided Bakewell's errors, has wrought a decided improvement in this breed. The improved Lincolns possess a rather more desirable robustness, approaching, in some few specimens, almost to coarseness, as compared with the finest Leicesters; but they are more hardy, and less liable to disease. They attain as large a size, and yield as great an amount of wool, of about the same value. This breed, indeed, scarcely differs more from the Cotswold than do flocks of a similar variety, which have been separately bred for several generations, from each other. They are prolific, and when well-fed, the ewes will frequently produce two lambs at a birth, for which they provide liberally from their udders till the time for weaning. The weight of the fleece varies from four to eight pounds per head.

Having alluded to the principal points of interest connected with the various breeds of sheep in the United States, our next business is with

---

THE NATURAL HISTORY OF THE SHEEP.

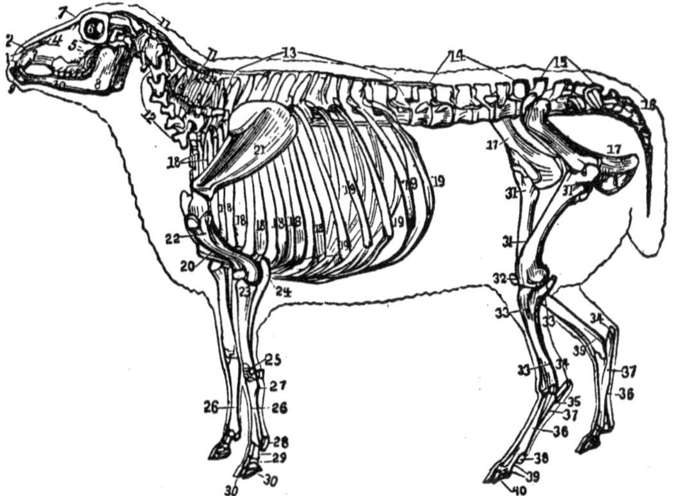

SKELETON OF THE SHEEP AS COVERED BY THE MUSCLES.

1. The intermaxillary bone. 2. The nasal bones. 3. The upper jaw 4 The union of the nasal and upper jaw-bones. 5. The union of the molar and lachrymal bones. 6. The orbits of the eye. 7. The frontal bone. 8. The lower jaw. 9 The incisor teeth, or nippers. 10. The molars or grinders. 11. The ligament of the neck supporting the head. 12. The seven vertebræ, or the bones of the neck. 13 The thirteen vertebræ, or bones of the back. 14. The six vertebræ of the loins. 15. The sacral bone.

16. The bones of the tail, varying in different breeds from twelve to twenty-one. 17. The haunch and pelvis. 18. The eight true ribs, with their cartilages. 19. The five false ribs, or those that are not attached to the breast-bone. 20. The breast-bone. 21. The scapula, or shoulder-blade. 22. The humerus, bone of the arm, or lower part of the shoulder. 23. The radius, or bone of the fore-arm. 24. The ulna or elbow. 25. The knee with its different bones. 26. The metacarpal or shank-bones—the larger bones of the leg. 27. A rudiment of the smaller metacarpal. 28. One of the sessamoid bones. 29. The first two bones of the foot—the pasterns. 30. The proper bones of the foot. 31. The thigh-bone. 32. The stifle-joint and its bone—the patella. 33. The tibia, or bone of the upper part of the leg. 34. The point of the hock. 35 The other bones of the hock. 36. The metatarsal bones, or bone of the hind-leg. 37. Rudiment of the small metatarsal. 38. A sessamoid bone. 39. The first two bones of the foot—the pasterns. 40. The proper bones of the foot.

DIVISION. *Vertebrata*—possessing a back-bone.

CLASS. *Mammalia*—such as give suck.

ORDER. *Ruminantia*—chewing the cud.

FAMILY. *Capridæ*—the goat kind.

GENUS. *Oris*—the sheep family.

Of this *Genus* there are three varieties:

ORIS, AMMON, or ARGALI.

Oris Musmon.

Oris Aries, or Domestic Sheep.

Of the latter—with which alone this treatise is concerned—there are about forty well known varieties. Between the *oris*, or sheep, and the *capra*, or goat, another *genus* of the same family, the distinctions are well marked, although considerable resemblance exists between them. The horns of the sheep have a spiral direction, while those of the goat have a direction upward and backward; the sheep, except in a single wild variety, has no beard, while the goat is bearded; the goat, in his highest state of improvement, when he is made to produce wool of a fineness unequalled by the sheep—as in the Cashmere breed—is mainly, and always, externally covered with hair, while the hair on the sheep may, by domestication, be reduced to a few coarse hairs, or got rid of altogether; and, finally, the pelt or skin of the goat has thickness very far exceeding that of the sheep.

The age of sheep is usually reckoned, not from the time that they are dropped, but from the first shearing; although the first year may thus include fifteen or sixteen months, and sometimes more. When doubt exists relative to the age, recourse is had to the teeth, since there is more uncertainty about the horn in this animal than in cattle; ewes that have been early bred, appearing always, according to the rings on the horn, a year older than others that have been longer kept from the ram.

FORMATION OF THE TEETH.

Sheep have no teeth in the upper jaw, but the bars or ridges of the palate thicken as they approach the forepart of the mouth; there also the dense, fibrous, elastic matter, of which they are constituted, becomes condensed, and forms a cushion or bed, which covers the converse extremity of the upper jaw, and occupies the place of the upper incisor, or cutting teeth, and partially discharge their functions. The herbage is firmly held between the front teeth in the lower jaw and this pad, and thus partly bitten and partly torn asunder. Of this, the rolling motion of the head is sufficient proof.

The teeth are the same in number as in the mouth of the ox. There are eight incisors or cutting-teeth in the forepart of the lower jaw, and six molars in each jaw above and below, and on either side. The incisors are more admirably formed for grazing than in the ox. The sheep lives closer, and is destined to follow the ox, and gather nourishment where that animal would be unable to crop a single blade. This close life not only loosens the roots of the grass, and disposes them to spread, but by cutting off the short suckers and sproutings—a wise provision of nature—causes the plants to throw out fresh, and more numerous, and stronger ones, and thus is instrumental in improving and increasing the value of the crop. Nothing will more expeditiously and more effectually make a thick, permanent pasture than its being occasionally and closely eaten down by sheep.

In order to enable the sheep to bite this close, the upper lip is deeply divided, and free from hair about the centre of it. The part of the tooth above the gum is not only, as in other animals, covered with enamel, to enable it to bear and to preserve a sharpened edge, but the enamel on the upper part rises from the bone of the tooth nearly a quarter of an inch, and presenting a convex surface outward, and a concave within, forms a little scoop or gorge of wonderful execution.

The mouth of the lamb newly dropped is either without incisor teeth or it has two. The teeth rapidly succeed to each other, and before the animal is a month old he has the whole of the eight. They continue to grow with his growth until he is about fourteen or sixteen months old. Then, with the same previous process of diminution as in cattle, or carried to a still greater degree, the two central teeth are shed, and attain their full growth when the sheep is two years old.

In examining a flock of sheep, however, there will often be very considerable difference in the teeth of those that have not been sheared, or those that have been once sheared; in some measure to be accounted for by a difference in the time of lambing, and likewise by the general health and vigor of the animal. There will also be a material difference in different animals, attributable to the good or bad keep which they have had. Those fed on good land, or otherwise well kept, will generally take the start of others that have been half starved, and renew their teeth some months sooner than these. There are also irregularities in the times of renewing the teeth, not to be accounted for by either of these circumstances; in fact, not to be explained by any known circumstance relating to the breed or the keep of the sheep. The want of improvement in sheep, which is occasionally observed, and which cannot be accounted for by any deficiency or change of food, may sometimes be justly attributed to the tenderness of the mouth when the permanent teeth are protruding through the gums.

Between two and three years old the next two incisors are shed; and when the sheep is actually three years old, the four central teeth are fully grown; at four years old, he has six teeth fully grown; and at five years old—one year before the horse or the ox can be said to be full-mouthed—all the teeth are perfectly developed. The sheep is a much shorter-lived animal than the horse, and does not often attain the usual age of the ox. Their natural age is about ten years, to which age they will breed and thrive well; though there are recorded instances of their breeding at the age of fifteen, and of living twenty years.

The careless examiner may be sometimes deceived with regard to the four-year-old mouth. He will see the teeth perfectly developed, no diminutive ones at the sides, and the mouth apparently full; and then, without giving himself the trouble of counting the teeth, he will conclude that the animal is five years old. A process of displacement, as well as of diminution, has taken place here; the remaining outside milk-teeth have not only shrunk to less than a fourth part of their original size, but the four-year-old teeth have grown before them and perfectly conceal them, unless the mouth is completely opened.

After the permanent teeth have all appeared and are fully grown, there is no criterion as to the age of the sheep In most cases, the teeth remain sound for one or two years, and then, at uncertain intervals—either on account of the hard work in which they have been employed, or from the natural effect of age—they begin to loosen and fall out; or, by reason of their natural slenderness, they are broken off. When favorite ewes, that have been kept for breeding, begin to lose condition, at six or seven years old, their mouths should be carefully examined. If any of the teeth are loose, they should be extracted, and a chance given to the animal to show how far, by browsing early and late, she may be able to make up for the diminished number of her incisors. It frequently happens that ewes with broken teeth, and some with all the incisors gone, will keep pace in condition with the best in the flock; but they must be well taken care of in the winter, and, indeed, nursed to an extent that would scarcely answer the farmer's purpose to adopt as a general rule, in order to prevent them from declining to such a degree as would make it very difficult afterward to fatten them for the butcher. It may certainly be taken as a general rule, that when sheep become broken-mouthed they begin to decline.

Causes of which the farmer is utterly ignorant, or over which he has no control, will sometimes hasten the loss of the teeth. One thing, however, is certain—that close feeding, causing additional exercise, does wear them down; and that the sheep of farmers who stock unusually and unseasonably hard, lose their teeth much sooner than others do.

THE STRUCTURE OF THE SKIN.

The skin of the sheep, in common with that of most animals, is composed of three textures. Externally is the cuticle, or scarf-skin, which is thin, tough, devoid of feeling, and pierced by innumerable minute holes, through which pass the fibres of the wool and the insensible perspiration. It seems to be of a scaly texture; although is not so evident as in many other animals, on account of a peculiar substance—the yolk—which is placed on it, to protect and nourish the roots of the wool. It is, however, sufficiently evident in the scab and other cutaneous eruptions to which this animal is liable.

Below this cuticle is the *rete mucosum*, a soft structure; its fibres having scarcely more consistence than mucilage, and being with great difficulty separated from the skin beneath. This appears to be placed as a defence to the terminations of the blood-vessels and nerves of the skin, which latter are, in a manner, enveloped and covered by it. The color of the skin, and probably that of the hair or wool also, is determined by the *rete mucosum;* or, at least, the hair and wool are of the same color as this substance.

Beneath the *retê mucosum* is the *cutis*, or true skin, composed of numberless minute fibres crossing each other in every direction; highly elastic, in order to fit closely to the parts beneath, and to yield to the various motions of the body; and dense and firm in its structure, that it may resist external injury. Blood-vessels and nerves innumerable pierce it, and appear on its surface in the form of *papillæ*, or minute eminences; while, through thousands of little orifices, the exhelant absorbents pour out the superfluous or redundant fluid. The true skin is composed, principally or almost entirely, of gelatine; so that, although it may be dissolved by long-continued boiling, it is insoluble in water at the common temperature. This organization seems to have been given to it, not only for the sake of its preservation while on the living animal, but that it may afterwards become useful to man. The substance of the hide readily combining with the tanning principle, is converted into leather.

THE ANATOMY OF THE WOOL.

On the skin of most animals is placed a covering of feathers, fur, hair, or wool. These are all essentially the same in composition, being composed of an animal substance resembling coagulated albumen, together with sulphur, silica, carbonate and phosphate of lime, and oxides of iron and manganese.

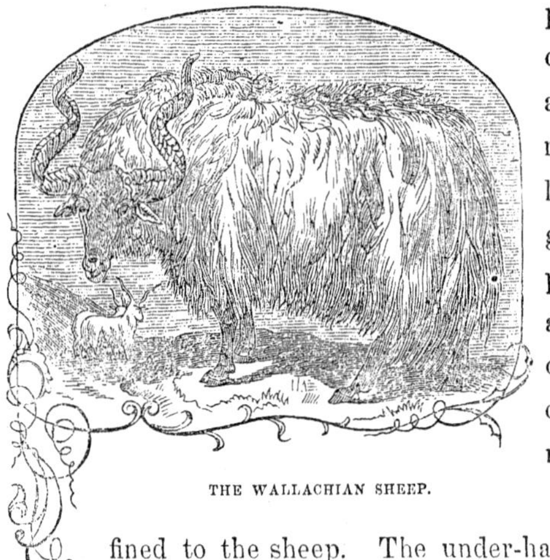

THE WALLACHIAN SHEEP.

Wool is not confined to the sheep. The under-hair of some goats is not only finer than the fleece of any sheep, but it occasionally has the crisped appearance of wool; being, in fact, wool of different qualities in different breeds—in some, rivalling or excelling that of the sheep, but in others very coarse. A portion of wool is also found on many other animals; as the deer, elk, the oxen of Tartary and Hudson's Bay, the gnu, the camel, many of the fur-clad animals, the sable, the polecat, and several species of the dog.

Judging from the mixture of wool and hair in the coat of most animals, and the relative situation of these materials, it is not improbable that such was the character of the fleece of the primitive sheep. It has, indeed, been asserted that the primitive sheep was entirely covered with hair; but this is, doubtless, incorrect. There exists, at the present day, varieties of the sheep occupying extensive districts, that are clothed outwardly with hair of different degrees of fineness and sleekness; and underneath the external coat is a softer, shorter, and closer one, that answers to the description of fur—according to most travellers—but which really possesses all the characteristics of wool. It is, therefore, highly improbable that the sheep—which has now become, by cultivation, the wool-bearing animal in a pre-eminent degree—should, in any country, at any time, have ever been entirely destitute of wool. Sheep of almost every variety have at times been in the gardens of the London (Eng.) Zoölogical Society; but there has not been one on which a portion of crisped wool, although exceedingly small, has not been discovered beneath the hair. In all the regions over which the patriarchs wandered, and extending northward through the greater part of Europe and Asia, the sheep is externally covered with hair; but underneath is a fine, short, downy wool, from which the hair is easily separated.

This is the case with the sheep at the Cape of Good Hope, and also in South America.

The change from hair to wool, though much influenced by temperature, has been chiefly effected by cultivation. Wherever hairy sheep are now found the management of the animal is in a most disgraceful state; and among the cultivated sheep the remains of this ancient hairy covering only exists, to any great extent, among those that are comparatively neglected or abandoned.

The filament of the wool has scarcely pushed itself through the pore of the skin, when it has to penetrate through another and singular substance, which, from its adhesiveness and color, is called *the yolk*. This is found in greatest quantity about the breast and shoulders—the very parts that produce the best, and healthiest, and most abundant wool—and in proportion as it extends, in any considerable degree, over other parts, the wool is then improved. It differs in quantity in different breeds. It is very abundant on the Merinos; it is sufficiently plentiful on most of the southern breeds, either to assist in the production of the wool, or to defend the sheep from the inclemency of the weather; but in the northern districts, where the cold is more intense and the yolk of wool is deficient, a substitute for it is sometimes sought by smearing the sheep with a mixture of tar, oil, or butter. Where there is a deficiency of yolk, the fibre of the wool is dry, harsh, and weak, and the whole fleece becomes thin and hairy; where the natural quantity of it is found, the wool is soft, oily, plentiful and strong.

This yolk is not the inspissated or thickened perspiration of the animal it is not composed of matter which has been accidentally picked up, and which has lodged in the wool; but it is a peculiar secretion from the glands of the skin, destined to be one of the agents in the nourishment of the wool, and at the same time, by its adhesiveness, to mat the wool together, and form a secure defence from the wet and cold.

Chemical experiments have established its composition, as follows: first, of a soapy matter with a basis of potash, which forms the greater part of it; second, a small quantity of carbonate of potash; third, a perceptible quantity of acetate of potash; fourth, lime, in a peculiar and unknown state of combination; fifth, an atom of muriate of potash; sixth, an animal oil, to which its peculiar odor is attributable. All these materials are believed to be essential to the yolk, and not found in it by mere accident, since the yolk of a great number of samples—Spanish, French, English, and American—has been subjected to repeated analyses, with the same result.

The yolk being a true soap, soluble in water, it is not difficult to account for the comparative ease with which sheep that have the natural proportion of it are washed in a running stream. There is, however, a small quantity of fatty matter in the fleece, which is not in combination with the alkali, and which, remaining attached to the wool, keeps it a little glutinous, notwithstanding the most careful washing.

The fibre of the wool having penetrated the skin and escaped from the yolk, is of a circular form, generally larger toward the extremity, and also toward the root, and in some instances very considerably so. The filaments of white wool, when cleansed from grease, are semi-transparent; their surface in some places is beautifully polished, in others curiously incrusted, and they reflect the rays of light in a very pleasing manner. When viewed by the aid of a powerful achromatic microscope, the central part of the fibre has a singularly glittering appearance. Minute filaments, placed very regularly, are sometimes seen branching from the main trunk, like boughs from the principal stem. This exterior polish varies much in different wools, and in wools from the same breed of sheep at different times. When the animal is in good condition, and the fleece healthy, the appearance of the fibre is really brilliant; but when the state of the constitution is bad, the fibre has a dull appearance, and either a wan, pale light, or sometimes scarcely any, is reflected. As a general rule, the filament is most transparent in the best and most useful wools, whether long or short. It increases with the improvement of the breed, and the fineness and healthiness of the fleece; yet it must be admitted that some wools have different degrees of the transparency and opacity, which do not appear to affect their value and utility. It is, however, the difference of transparency in the same fleece, or in the same filament, that is chiefly to be noticed as improving the value of the wool.

As to the size of the fibre, the terms "fine" and "coarse," as commonly used, are but vague and general descriptions of wool. All fine fleeces have some coarse wool, and all coarse fleeces some fine. The most accurate classification is to distinguish the various qualities of wool in the order in which they are esteemed and preferred by the manufacturer—as the following: first, fineness with close ground, that is, thick-matted ground; second, pureness; third, straight-haired, when broken by drawing; fourth, elasticity, rising after compression in the hand; fifth, staple not too long; sixth, color; seventh, what coarse exists to be very coarse; eighth, tenacity; and ninth, not much pitch-mark, though this is no disadvantage, except the loss of weight in scouring. The bad or disagreeable properties are—thin, grounded, tossy, curly-haired, and, if in a sorted state, little in it that is very fine; a tender staple, as elasticity, many dead white hairs, very yolky. Those who buy wool for combing and other light goods that do not need milling, wish to find length of staple, fineness of hair, whiteness, tenacity, pureness, elasticity, and not too many pitch-marks.

The property first attracting attention, and being of greater importance than any other, is *the fineness* of the pile—the quantity of fine wool which a fleece yields, and the degree of that fineness. Of the absolute fineness, little can be said, varying, as it does, in different parts of the same fleece to a very considerable degree, and the diameter of the same fibre often being exceedingly different at the extremity and the centre. The micrometer has sometimes indicated that the diameter of the former is five times as much as that of the latter; and, consequently, that a given length of yield taken from the extremity would weigh twenty-five times as much as the same length taken from the centre and cleansed from all yolk and grease. That fibre may be considered as coarse whose diameter is more than the five-hundredth part of an inch; in some of the most valuable samples of Saxony wool it has not exceeded the nine-hundredth part; yet in some animals, whose wool has not been used for manufacturing purposes, it is less than one twelve-hundredth part.

The extremities of the wool, and frequently those portions which are near to the root, are larger than the intermediate parts. The extremity of the fibre has, generally, the greatest bulk of all. It is the product of summer, soon after shearing-time, when the secretion of the matter of the wool is increased, and when the pores of the skin are relaxed and open, and permit a larger fibre to protrude. The portion near the root is the growth of spring, when the weather is getting warm; and the intermediate part is the offspring of winter, when under the influence of the cold the pores of the skin contract, and permit only a finer hair to escape. If, however, the animal is well fed, the diminution of the bulk of the fibre will not be followed by weakness or decay, but, in proportion as the pile becomes fine, the value of the fleece will be increased; whereas, if cold and starvation should go hand-in-hand, the woolly fibre will not only diminish in bulk, but in health, strength, and worth.

The variations in the diameter of the wool in different parts of the fibre will also curiously correspond with the degree of heat at the time the respective portions were produced. The fibre of the wool and the record of the meteorologist will singularly agree, if the variations in temperature are sufficiently distinct from each other for any appreciable part of the fibre to form. It follows from this, that—the natural tendency to produce wool of a certain fibre being the same—sheep in a hot climate will yield a comparatively coarse wool, and those in a cold climate will carry a finer, but at the same time a closer and a warmer fleece. In proportion to the coarseness of a fleece will generally be its openness, and its inability to resist either cold or wet; while the coat of softer, smaller, more pliable wool will admit of no interstices between its fibres, and will bid defiance to frost and storms.

The natural instinct of the sheep would seem to teach the

wool-grower the advantage of attending to the influence of temperature upon the animal. He is evidently impatient of heat. In the open districts, and where no shelter is near, he climbs to the highest parts of his walk, that, if the rays of the sun must still fall on him, he may nevertheless be cooled by the breeze; but, if shelter is near, of whatever kind, every shaded spot is crowded with sheep. The wool of the Merinos after shearing-time is hard and coarse to such a degree as to render it very difficult to suppose that the same animal could bear wool so opposite in quality, compared with that which had been clipped from it in the course of the same season. As the cold weather advances, the fleeces recover their soft quality.

Pasture has a far greater influence on the fineness of the fleece. The staple of the wool, like every other part of the sheep, must increase in length or in bulk when the animal has a superabundance of nutriment; and, on the other hand, the secretion which forms the wool must decrease like every other, when sufficient nourishment is not afforded. When little cold has been experienced in the winter, and vegetation has scarcely been checked, the sheep yields an abundant crop of wool, but the fleece is perceptibly coarser as well as heavier. When the frost has been severe, and the ground long covered with snow, if the flock has been fairly supplied with nutriment, although the fleece may have lost a little in weight, it will have acquired a superior degree of fineness, and a proportional increase of value. Should, however, the sheep have been neglected and starved during this continued cold weather, the fleece as well as the carcass is thinner, and although it may have preserved its smallness of filament, it has lost in weight, and strength, and usefulness.

Connected with fineness is *trueness of staple*—as equal in growth as possible over the animals—a freedom from those shaggy portions, here and there, which are occasionally observed on poor and neglected sheep. These portions are always coarse and comparatively worthless, and they indicate an irregular and unhealthy action of the secretion of wool, which will also probably weaken or render the fibre diseased in other parts. Included in trueness of fibre is another circumstance to which allusion has already been made—a freedom from coarse hairs which project above the general level of the wool in various parts, or, if they are not externally seen, mingle with the wool and debase its qualities.

Soundness is closely associated with trueness. It means, generally speaking, strength of the fibre, and also a freedom from those breaches or withered portions of which something has previously been said. The eye will readily detect the breaches; but the hair generally may not possess a degree of strength proportioned to its bulk. This is ascertained by drawing a few hairs out of the staple, and grasping each of them singly by both ends, and pulling them until they break. The wool often becomes injured by felting while it is on the sheep's back. This is principally seen in the heavy breeds, especially those that are neglected and half-starved, and generally begins in the winter season, when the coat has been completely saturated with water, and it increases until shearing-time, unless the cob separates from the wool beneath, and drops off.

Wool is generally injured by keeping. It will probably increase a little in weight for a few months, especially if kept in a damp place; but after that it will somewhat rapidly become lighter, until a very considerable loss will often be sustained. This, however, is not the moral of the case; for, except very great care is taken, the moth will get into the bundles and injure and destroy the staple; and that which remains untouched by them will become considerably harsh and less pliable. If to this the loss of the interest of money is added, it will be seen that he seldom acts wisely who hoards his wool, when he can obtain what approaches to a fair remunerating price for it.

Softness of the wool is evidently connected with the presence and quality of the yolk. This substance is undoubtedly designed not only to nourish the hair, but to give it richness and pliability. The growth of the yolk ought to be promoted, and agriculturists ought to pay more attention to the quantity and quality of yolk possessed by the animals selected for the purpose of breeding.

Bad management impairs the pliability of the wool, by arresting the secretion of the yolk. The softness of the wool is also much influenced by the chemical elements of the soil. A chalky soil notoriously deteriorates it; minute particles of the chalk being necessarily brought into contact with the fleece and mixing with it, have a corrosive effect on the fibre, and harden it and render it less pliable. The particles of chalk come in contact with the yolk—there being a chemical affinity between the alkali and the oily matter of the yolk—immediately unite, and a true soap is formed. The first storm washes a portion of it; and the wool, deprived of its natural pabulum and unguent, loses some of its vital properties—its pliability among the rest. The slight degree of harshness which has been attributed to the English South-Down has been explained in this way.

The felting property of wool is a tendency of the fibres to entangle themselves together, and to form a mass more or less difficult to unravel. By moisture and pressure, the fibres of the wool may become matted or felted together into a species of cloth. The manufacture of felt was the first mode in which wool was applied to clothing, and felt has long been in universal use for hats. The fulling of flannels and broadcloths is effected by the felting principle. By the joint influence of the moisture and the pressure, certain of the fibres are brought into more intimate contact with each other; they adhere—not only the fibres, but, in a manner, the threads—and the cloth is taken from the mill shortened in all its dimensions; it has become a kind of felt, for the threads have disappeared, and it can be cut in every direction with very little or no unravelling; it is altogether a thicker, warmer, softer fibre. This felting property is one of the most valuable qualities possessed by wool, and on this property are the finer kinds of wool especially valued by the manufacturer for the finest broadcloths. This naturally suggests a consideration of the various forms in the structure on which it depends.

The most evident distinction between the qualities of hair and wool is the comparative straightness of the former, and *the crisped or spirally-curling form* which the latter assumes. If a little lock of wool is held up to the light, every fibre of it is twisted into numerous minute corkscrew-like ringlets. This is especially seen in the fleece of the short-woolled sheeps; but, although less striking, it is obvious even in wool of the largest staple.

The spirally-curving form of wool used, erroneously, to be considered as the chief distinction between the covering of the goat and the sheep; but the under-coat of some of the former is finer than that of any sheep, and it is now acknowledged frequently to have the crisped and curled appearance of wool. In some breeds of cattle, particularly in one variety of the Devons, the hair assumes a curled and wavy appearance, and a few of the minute spiral ringlets have been occasionally seen. It is the same with many of the Highlands; but there is no determination to take on the true crisped character, and throughout its whole extent, and it is still nothing but hair. On some foreign breeds, however, as the yak of Tartary, and the ox of Hudson's Bay, some fine and valuable wool is produced.

There is an intimate connection between the fineness of the wool and the number of the curves, at least in sheep yielding wool of nearly the same length; so that, whether the wool of different sheep is examined, or that from different parts of the same sheep, it is enough for the observer to take advice of the number of curves in a given space, in order to ascertain with sufficient accuracy the fineness of the fibre.

To this curled form of the wool not enough attention is, as a general thing, paid by the breeder. It is, however, that on which its most valuable uses depend. It is that which is essential to it in the manufactory of cloths. The object of the carder is to break the wool in pieces at the curves—the principle of the thread is the adhesion of the particles together by their curves; and the fineness of the thread, and consequent fineness of the cloth, will depend on the minuteness of these curves, or the number of them found in a given length of fibre.

It will readily be seen that this curling form has much to do with the felting property of wool; it materially contributes to that disposition in the fibres which enables them to attach and intwine themselves together; it multiplies the opportunities for this interlacing, and it increases the difficulty of unravelling the felt.

The felting property of wool is the most important, as well as the distinguishing one; but it varies essentially in different breeds, and the usefulness and the consequent value of the fleece, for clothing purposes, at least, depend on the degree to which it is pursued.

The serrated—notched, like the teeth of a saw—*edge* of wool, which has been discovered by means of the microscope, is also, as well as the spiral curl, deemed an important quality in the felting property. Repeated microscopic observations have removed all doubts as to the general outline of the woolly fibre. It consists of a central stem or stalk, probably hollow, or, at least, porous, possessing a semi-transparency, not found in the fibre of hair. From this central stalk there springs, at different distances, on different breeds of sheep, a circlet of leaf-shaped projections.

LONG WOOL.

The most valuable of the long-woolled fleeces are of British origin. A considerable quantity is produced in France and Belgium; but the manufacturers in those countries acknowledge the superiority of the British wool. Long wool is distinguished, as its name would import, by the length of its staple, the average of which is about eight inches. It was much improved, of late years, both in England and in other countries. Its staple has, without detriment to its manufacturing qualities, become shorter; but it has also become finer, truer, and sounder. The long-woolled sheep has been improved more than any other breed; and the principal error which Bakewell committed having been repaired since his death, the long wool has progressively risen in value, at least for curling purposes. Some of the breeds have staples of double the length that has been mentioned as the average one. Pasture and breeding are the powerful agents here.

Probably because the Leicester blood prevails in, or, at least, mingles with, every other long-woolled breed, a great similarity in the appearance and quality of this fleece has become apparent, of late years, in every district of England. The short-woolled fleeces are, to a very considerable degree, unlike in fineness, elasticity, and felting property; the sheep themselves are still more unlike; but the long-wools have, in a great degree, lost their distinctive points—the Lincoln, for example, has not all of his former gaunt carcass, and coarse, entangled wool—the Cotswold has become a variety of the Leicester—in fact, all the long-woolled sheep, both in appearance and fleece, have almost become of one variety; and rarely, except from culpable neglect in the breeder, has the fleece been injuriously weakened, or too much shortened, for the most valuable purposes to which it is devoted.

In addition to its length, this wool is characterized by its strength, its transparency, its comparative stoutness, and the slight degree in which it possesses the felting property. Since the extension of the process of combing to wools of a shorter staple, the application of this wool to manufacturing purposes has undergone considerable change. In some respects, the range of its use has been limited; but its demand has, on the whole, increased, and its value is more highly appreciated. Indeed, there are certain important branches of the woollen manufacture, such as worsted stuffs, bombazines, muslin-de-laines, etc., in which it can never be superseded; and its rapid extension in the United States, within the past few years, clearly shows that a large and increasing demand for this kind of wool will continue at remunerating prices.

This long wool is classed under two divisions, distinguished both by length and the fineness of the fibre. The first—*the long-combing wool*—is used for the manufacture of hard yarn, and the worsted goods for which that thread is adapted, and requires the staple to be long, firm, and little disposed to felt. *The short-combing wool* has, as its name implies, a shorter staple, and is finer and more felty; the felt is also closer and softer, and is chiefly used for hosiery goods.

MIDDLE WOOL.

This article is of more recent origin than the former, but has rapidly increased in quantity and value. It can never supersede, but will only stand next in estimation to, the native English long fleece. It is yielded by the half-bred sheep—a race that becomes more numerous every year—being a cross of the Leicester ram with the South-Down, or some other short-woolled ewe; retaining the fattening property and the early maturity of the Leicester, or of both; and the wool deriving length and straightness of fibre from the one, and fineness and feltiness from the other. The average length of staple is about five inches. There is no description of the finer stuff-goods in which this wool is not most extensively and advantageously employed; and the nails, or portions which are broken off by the comb, and left in, whether belonging to this description of wool or to the long wool, are used in the manufacture of several species of cloth of no inferior quality or value.

Under the breed of middle wools must be classed those which, when there were but two divisions, were known by the name of short wools; and if English productions were alone treated of, would still retain the same distinctive appellation. To this class belong the South-Down and Cheviot; together with the fleece of several other breeds, not so numerous, nor occupying so great an extent of country. From the change, however, which insensibly took place in them all—the lengthening, and the increased thickness of the fibre, and, more especially, from the gradual introduction of other wools possessing delicacy of fibre, pliability, and felting qualities beyond what these could claim, and at the same time, being cheaper in the market—they lost ground in the manufacture of the finer cloths, and have for some time ceased to be used in the production of them. On the other hand, the changes which have taken place in the construction of machinery have multiplied the purposes to which they may be devoted, and very considerably enhanced their value.

These wools, of late, rank among the combing wools; they are prepared as much by the comb as by the card, and in some places more. On this account they meet with a readier sale, at fair, remunerating prices, considering the increased weight of each individual fleece, and the increased weight and earlier maturity of the carcass. The South-Downs yield about seven-tenths of the pure short wools grown in the British kingdoms; but the half-bred sheep has, as has been remarked, encroached on the pure short-woolled one. The average staple of middle-woolled sheep is three and a half inches.

These wools are employed in the manufacture of flannels, army and navy cloths, coatings, heavy cloths for calico printers and paper manufacturers, woollen cords, coarse woollens, and blankets; besides being partially used in cassinettes, baizes, bockings, carpets, druggets, etc.

SHORT WOOL.

From this division every wool of English production is excluded. These wools, yielded by the Merinos, are employed, unmixed, in the manufacture of the finer cloths, and, combined with a small proportion of wool from the English breeds, in others of an inferior value. The average length of staple is about two and a half inches.

These wools even may be submitted to the action of the comb. There may be fibres only one inch in length; but if there are others from two and a half to three inches, so that the average of the staple shall be two inches, a thread sufficiently tenacious may, from the improved state of machinery, be spun, and many delicate and beautiful fabrics readily woven, which were unknown not many years ago.

No one breed of sheep combines the highest perfection in all those points which give value to this race of animals. One is remarkable for the weight, or early maturity, or excellent quality of its carcass, while it is deficient in quality or quantity of wool; and another, which is valuable for wool, is comparatively deficient in carcass. Some varieties will flourish only under certain conditions of food and climate; while others are much less affected by those conditions, and will subsist under the greatest variations of temperature, and on the most opposite qualities of verdure.

In selecting a breed for any given locality, reference should be had, *first*, to the feed and climate, or the surrounding natural circumstances; and, *second*, to the market facilities and demand. Choice should then be made of that breed which, with the advantages possessed, and under all the circumstances, will yield the greatest net value of the marketable product.

Rich lowland herbage, in a climate which allows it to remain green during a large portion of the year, is favorable to the production of large carcasses. If convenient to a market where mutton finds a prompt sale and good prices, then all the conditions are realized which calls for a mutton-producing, as contradistinguished from a wool-yielding, sheep. Under such circumstances, the choice should undoubtedly be made from the improved English varieties—the South-Down, the New Leicester, and the improved Cotswolds or New Oxfordshire sheep. In deciding between these, minor and more specific circumstances must be taken into account. If large numbers are to be kept, the Downs will herd—remain thriving and healthy when kept together in large numbers—much better than the two larger breeds; if the feed, though generally plentiful, is liable to be somewhat short during the droughts of summer, and there is not a certain supply of the most nutritious winter feed, the Downs will better endure occasional short keep; if the market demands a choice and high-flavored mutton, the Downs possess a decided superiority. If, on the other hand, but few are to be kept in the same enclosure, the large breeds will be as healthy as the Downs; if the pastures are somewhat wet or marshy, the former will better subsist on the rank herbage which usually grows in such situations; if they do not afford so fine a quality of mutton, they—particularly the Leicester—possess an earlier maturity, and give more meat for the amount of food consumed, as well as yield more tallow.

The next point of comparison between the long and the middle woolled families, is the value of their wool. Though not the first or principal object aimed at in the cultivation of any of these breeds, it is, in this country, an important item or incident in determining their relative profitableness. The American Leicester yields about six pounds of long, coarse, combing wool; the Cotswold, somewhat more; but this perhaps counterbalanced by these considerations; the Downs grow three to four pounds of a low quality of carding wool. None of these wools are very salable, at remunerating prices, in the American markets. Both, however, will appreciate in proportion to the increase of manufactures of worsted, flannels, baizes, and the like. The difference in the weight of the fleeces between the breeds is, of itself, a less important consideration than it would at first appear, for reasons which will be given when the connection between the amount of wool produced and the food consumed by the sheep is noticed.

The Cheviots are unquestionably inferior to the breeds above named, except in a capacity to endure a vigorous winter, and to subsist on healthy herbage. Used in the natural and artificial circumstances which surround sheep-husbandry in many parts of England—where the fattest and finest quality of mutton is consumed, as almost the only animal food of the laboring classes—the heavy, early-maturing New Leicester, and the still heavier New Oxfordshire sheep seem exactly adapted to the wants of producers and consumers, and are of unrivalled value. To depasture poorer soils, sustain a folding system, and furnish the mutton which supplies the tables of the wealthy, the South-Down meets an equal requirement.

Sufficient attention is by no means paid in many portions of the country to the profit which could be made to result from the cultivation of the sheep. One of the most serious defects in the prevalent husbandry of New England, for example, is the neglect of sheep. Ten times the present number might be easily fed, and they would give in meat, wool, and progeny, more direct profit than any other domestic animal, while the food which they consume would do more towards fertilizing the farms than an equal amount consumed by any other animal. It is notorious that the pastures of that section of the country have seriously deteriorated in fertility and become overrun with worthless weeds and bushes to the exclusion of nutritious grasses.

With sheep—as well as with all other animals—much or prolonged exercise in pursuit of food, or otherwise, is unfavorable to taking on fat. Some seem to forget, in their earnest advocacy of the merits of the different breeds, that the general physical laws which control the development of all the animal tissues as well as functions, are uniform. Better organs will, doubtless, make a better appropriation of animal food; and they may be taught, so to speak, to appropriate it in particular directions: in one breed, more especially to the production of fat; in another, of muck, or lean meat; in yet another, of wool. But, these things being equal, large animals will always require more food than small ones. Animals which are to be carried to a high state of fatness must have plentiful and nutritious food, and they must exercise but little, in order to prevent the unnecessary combustion in the lungs of that carbon which forms nearly four-fifths of their fat. No art of breeding can counteract these established laws of Nature.

In instituting a comparison between breeds of sheep for *wool-growing* purposes, it is undeniable that the question is not, what variety will shear the heaviest or even the most valuable fleece, irrespective of the cost of production. Cost of feed and care, and every other expense, must be deducted, in order to fairly test the profits of an animal. If a large sheep consume twice as much food as a small one, and give but once and a-half as much wool, it is obviously more profitable—other things being equal—to keep two of the smaller sheep. The next question, then, is,—*from what breed*—with the same expense in other particulars—*will the verdure of an acre of land produce the greatest value of wool?*

And, first, as to the comparative amount of food consumed by the several breeds. There are no satisfactory experiments which show that *breed,* in itself considered, has any particular influence on the quantity of food consumed. It is found, with all varieties, that the consumption is in proportion to the live weight of the grown animal. Of course, this rule is not invariable in its individual application; but its general soundness has been satisfactorily established. Grown sheep take up between two and a half and three and a third per cent. of their weight, in what is equivalent to dry hay, to keep themselves in store condition.

The consumption of food, then, being proportioned to the weight, it follows that, if one acre is capable of sustaining three Merinos, weighing one hundred pounds each, it will sustain two Leicesters, weighing one hundred and fifty each, and two and two-fifth South-Downs, weighing one hundred and twenty-five each. Merinos of this weight often shear five pounds per fleece, taking flocks through. The herbage of an acre, then, would give fifteen pounds of Merino wool, twelve of Leicester, and but nine and three-fifths of South-Down—estimating the latter as high as four pounds to the fleece. Even the finest and lightest-fleeced sheep known as Merinos average about four pounds to the fleece; so that the feed of an acre would produce as much of the highest quality of wool sold under the name of Merino as it would of New Leicester, and more than it would of South-Down, while the former would be worth from fifty to one hundred per cent. more per pound than either of the latter.

Nor does this indicate all the actual difference, as in the foregoing estimate the live weight of the English breeds is placed low, and that of the Merinos high. The live weight of the five-pound fine-fleeced Merino does not exceed ninety pounds; it ranges, in fact, from eighty to ninety; so that three hundred pounds of live weight—it being understood that all of these live weights refer to ewes in fair ordinary, or what is called store, condition—would give a still greater product of wool to the acre. It is perfectly safe, therefore, to say that the herbage of an acre will uniformly give nearly double the value of Merino that it will of any of the English long or middle wools.

What are the other relative expenses of these breeds? The full-blooded Leicester is in no respect a hardier sheep than the Merino, though some of its crosses are much hardier than the pure-bred sheep: indeed, it is less hardy, under the most favorable circumstances. It is more subject to colds; its constitution more readily gives way under disease; the lambs are more liable to perish from exposure to cold, when newly dropped. Under unfavorable circumstances—herded in large flocks, famished for feed, or subjected to long journeys—its capacity to endure, and its ability to rally from sad drawbacks, do not compare with those of the Merino. The high-bred South-Down, though considerably less hardy than the unimproved parent stock, is still fairly entitled to the appellation of a hardy animal; it is, in fact, about on a pace with the Merino, though it will not bear as hard stocking, without a rapid diminution in size and quality. If the peculiar merits of the animal are to be considered in determining the expenses, as they surely should be, the superior fecundity of the South-Down is a point in its favor, as well for a wool-producing as a mutton sheep. The ewe not only frequently produces twin lambs—as do both the Merino and Leicester—but, unlike the latter, she possesses nursing properties to do justice to them. This advantage, however, is fully counterbalanced by the superior longevity of the Merino. All the English mutton breeds begin to rapidly deteriorate in amount of wool, capacity to fatten, and general vigor, at about five years old; and their early maturity is no offset to this, in an animal kept for wool-growing purposes. This early decay requires earlier and more rapid slaughter than is always economically convenient, or even possible.

It is well, on properly stocked farms, to slaughter or turn off the Merino wether at four or five years old, to make room for the breeding stock; but he will not particularly deteriorate, and he will richly pay the way with his fleece for several years longer. Breeding ewes are rarely turned off before eight, and are frequently kept until ten years old, at which period they exhibit no greater marks of age than do the Downs and Leicester at five or six. Instances are known of Merino ewes breeding uniformly until fifteen years old. The improved Cotswold is said to be hardier than the Leicester; but this variety, from their great size, and the consequent amount of food consumed by them, together with the other necessary incidents connected with the breeding of such large animals, is incapacitated from being generally introduced as a wool-growing sheep. All the coarse races have one advantage over the Merino: they are less subject to the visitation of the hoof-ail, and when untreated, this disease spreads with less violence and malignity among them. This has been explained by the fact that their hoofs do not grow long and turn under from the sides, as do those of the Merino, and thus retain dirt and filth in constant contact with the foot.

Taking into account all the circumstances connected with the peculiar management of each race, together with all the incidents, exigencies, and risks of the husbandry of each, it may be confidently asserted that the expenses, other than those of feed, are not smaller per head, or even in the number required to stock an acre, in either of the English breeds above referred to, than in the Merino. Indeed, it may well be doubted whether any of those English breeds, except the South-Down, is on an equality, even, with the Merino, in these respects. For wool-growing purposes, the Merino, then, possesses a marked and decided superiority over the best breeds and families of coarse-woolled sheep. As a mutton sheep, it is inferior to some of those breeds; although not so much as is popularly supposed. Many persons, who have never tasted Merino mutton, and who have, consequently, an unfavorable impression of it, would, if required to consume the fat and lean together, find it more palatable than the luscious and over-fat New Leicester. The mutton of the cross between the Merino and the Native would certainly be preferred to the Leicester, by anybody but an English laborer, accustomed to the latter, since it is short-grained, tender, and of good flavor. The same is true of the crosses with the English varieties, which will hereafter be treated of more particularly. Grade Merino wethers, half-bloods, for example, are favorites with the drover and butcher, being of good size, extraordinarily heavy for their apparent bulk, by reason of the shortness of their wool, compared with the coarse breeds, making good mutton, tallowing well, and their pelts, from the greater weight of wool on them, commanding an extra price. In speaking of the Merino in this connection, no reference is made to the Saxons, though they are, as is well known, pure-blooded descendants of the former.

Assuming it, then, as settled, that it is to the Merino race that the wool-grower must look for the most profitable sheep, a few considerations are subjoined as to the adaptability of the widely diverse sub-varieties of the race to the wants and circumstances of different portions of the country.

Upon the first introduction of the Saxons, they were sought with avidity by the holders of the fine-woolled flocks of the country, consisting at that time of pure or grade Merinos. Under the decisive encouragement offered both to the wool-grower and the manufacturer by the tariff of 1828, a great impetus was given to the production of the finest wools, and the Saxon everywhere superseded, or bred out by crossing, the Spanish Merinos. In New York and New England, the latter almost entirely disappeared. In the fine-wool mania which ensued, weight of fleece, constitution, and every thing else, were sacrificed to the quality of the wool. Then came the tariff of 1832, which, as well as that of 1828, gave too much protection to both wool-grower and manufacturer, into whose pursuits agricultural and mercantile speculators madly rushed. Skill without capital, capital without skill, and in some cases, probably, thirst for gain without either, laid hold of these favored avocations. The natural and inevitable result followed. In the financial crisis of 1837, manufacturing, and all other monetary enterprises which had not been conducted with skill and providence, and which were not based on an adequate and vast capital, were involved in a common destruction; and even the most solid and best conducted institutions of the country were shaken by the fury of the explosion. Wool suddenly fell almost fifty per cent. The grower began to be discouraged. The breeder of the delicate Saxons—and they comprised the flocks of nearly all the large wool-growers in the country, at that time—could not obtain for his wool its actual first cost per pound.

When the Saxon growers found that the tariff of 1842 brought them no relief, they began to give up their costly and carefully nursed flocks. The example once set, it became contagious; and then was a period when it seemed as if all the Saxon sheep of the country would be sacrificed to this reaction. Many abandoned wool-growing altogether, at a heavy sacrifice of their fixtures for rearing sheep; others crossed with coarse-woolled breeds; and, rushing from one extreme to the other, some even crossed with the English mutton breeds; or some, with more judgment, went back to the parent Merino stock, but usually selected the heaviest and coarsest-woolled Merinos, and thus materially deteriorated the character of their wool. This period became distinguished by a mania for heavy fleeces. The English crosses were, however, speedily abandoned. The Merino regained his supremacy, lost for nearly a quarter of a century, and again became the popular favorite. It was generally adopted by those who were commencing flocks in the new Western States, and gives its type to the sheep of those regions.

The supply of fine wool, then, proportionably decreased, and that of medium and coarse increased. Wools, for convenience, may be classified as follows: *superfine*, the choicest quality grown in the United States, and never grown here excepting in comparatively small quantities; *fine*, good ordinary Saxon; *good medium*, the highest quality of wool usually known in the market as Merino; *medium*, ordinary Merino; *ordinary*, grade Merino and selected South-Down fleeces; and, *coarse*, the English long-wools, etc. This subdivision is, perhaps, minute enough for all practical purposes here.

It soon became apparent that, to sustain our manufacturing interest—that engaged in the manufacture of fine cloths—the diminution of fine wools should not only be at once arrested, but that the growth of them should be immediately and largely increased. An increased attention was accordingly bestowed upon this branch of industry, and sections of the country which had previously held aloof from wool-growing, embarked in that calling with commendable enterprise.

The climate north of forty-one degrees, or, beyond all dispute, north of forty-two degrees, is too severe for any variety of sheep commonly known, which bear either superfine or fine wools. In fact, the only such variety in any thing like general use is the Saxon; and this, as has been remarked, is a delicate sheep, entirely incapable of safely withstanding our northern winters, without good shelter, good and regularly-administered food, and careful and skilful management in all other particulars. When the season is a little more than usually back-hand, so that grass does not start prior to the lambing season, it is difficult to raise the lambs of the mature ewes; the young ewes will, in many instances, disown their lambs, or, if they own them, not have a drop of milk for them; and if, under such circumstances, as often happens, a northeast or a north-west storm comes driving down, bearing snow or sleet in its wings, or there is a sudden depression of the temperature from any cause, no care will save multitudes of lambs from perishing. If the time of having the lambs dropped is deferred, for the purpose of escaping these evils, they will not attain size and strength sufficient to enable them to pass safely through their first winter. North of the latitude last named, it is necessary, as a general rule, that they be dropped in the first half of May, to give them this requisite size and strength; and occasional cold storms come, nearly every season, up to that period, and, not unfrequently, up to the first of June.

These considerations have had their weight even with the few large sheepholders in that section, whose farms and buildings have been arranged with exclusive reference to the rearing of these sheep; many of whom have adopted a Merino cross. With the ordinary farmers—the small sheep-owners, who, in the aggregate, grow by far the largest portion of the northern wools—the Saxon sheep is, for these reasons, in marked disrepute They have not the necessary fixtures for their winter protection, and are unwilling to bestow the necessary amount of care on them. Besides, mutton and wool being about an equal consideration with this class, they want larger and earlier maturing breeds. Above all, they want a strong, hardy sheep, which demands no more care than their cattle. The strong, compact, medium-woolled Merino, or, more generally, its crosses with coarse varieties, producing the wool classed as ordinary, is the common favorite. In the Northwest, this is especially the case, where the climate is still worse for delicate sheep.

At the South, on the contrary—where these disadvantages do not exist to so great an extent, certainly—wool varying from good medium upward are more profitable staples for cultivation than the lower classes; and in that section a high degree of fineness in fleece has been sought in breeding the Merino—the four-pound fine-fleeced Merino having received marked attention. This is a far more profitable animal than the Saxon, other things being equal—which is not the case, since the former is every way a hardier animal and a better nurse; and, although about twenty pounds heavier, and therefore consuming more feed, this additional expense is more than counterbalanced by the additional care and risk attending the husbandry of the Saxon.

POINTS OF THE MERINO.

For breeding purposes, the shape and general appearance of the Merino should be as follows:—The head should be well carried up, and in the ewe hornless. It would be better, on many accounts, to have the ram also hornless, but, as horns are usually characteristic of the Merino ram, many prefer to see them. The face should be rather short, broad between the eyes, the nose pointed, and, in the ewe, fine and free from wrinkles. The eye should be bright, moderately prominent, and gentle in its expression. The neck should be straight—not curving downward—short, round, and stout—particularly so at its junction with the shoulder, forward of the upper point of which it should not sink below the level of the back. The points of the shoulder should not rise to any perceptible extent above the level of the back. The back, to the hips, should be straight; the crops—that portion of the body immediately back of the shoulder-blades—full; the ribs well arched; the body large and capacious; the flank well let down; the hind-quarters full and round—the flesh meeting well down between the thighs, or in the "twists." The bosom should be broad and full; the legs short, well apart, and perpendicular—that is, not drawn under the body toward each other when the sheep is standing. Viewed as a whole, the Merino should present the appearance of a low, stout, plump, and—though differing essentially from the English mutton-sheep model—a highly symmetrical sheep.

The skin is an important point. It should be loose, singularly mellow, and of a rich, delicate pink color. A colorless skin, or one of a tawny, approaching to a butternut, hue, indicates bad breeding. On the subject of wrinkles, there is a difference of opinion. As they are rather characteristic of the Merino—like the black color in a Berkshire hog, or the absence of all color in Durham cattle—these wrinkles have been more regarded, by novices, than those points which give actual value to the animal; and shrewd breeders have not been slow to act upon this hint. Many have contended that more wool can be obtained from a wrinkled skin; and this view of the case has led both the Spanish and French breeders to cultivate them largely—the latter, to a monstrosity. An exceedingly wrinkled neck, however, adds but little to the weight of the fleece—not enough, in fact, to compensate for the deformity, and the great impediment thus placed in the way of the shearer. A smoothly drawn skin, and the absence of all dead lap, would not, on the other hand, perhaps be desirable.

The wool should densely cover the whole body, where it can possibly grow—from a point between and a little below the eyes, and well up on the cheeks, to the knees and hocks. Short wool may show, particularly in young animals, on the legs, even below the knees and hocks; but long wool covering the legs, and on the nose, below the eyes, is unsightly, without value; while on the face it frequently impedes the sight of the animal, causing it to be in a state of perpetual alarm, and disqualifying it to escape real danger. Neither is this useless wool the slightest indication of a heavy fleece—contrary to what seems to be thought by some. It is very often seen in Saxons shearing scarcely two pounds of wool, and on the very lightest fleeced Merinos.

The amount of gum which the wool should exhibit is another mooted point. Merino wool should be yolky, or oily, prior to washing—though not to the extreme extent, occasionally witnessed, of giving it the appearance of being saturated with grease. The extreme tips may exhibit a sufficient trace of gum to give the fleece a darkish cast, particularly in the ram; but a black, pitchy gum, resembling half-hardened tar, extending an eighth or a quarter of an inch into the fleece, and which cannot be removed by ordinary washing, is decidedly objectionable. There is a white or yellowish concrete gum, not removable by common washing, which appears in the interior of some fleeces, and is equally objectionable.

The weight of fleece remaining the same, medium length of staple, with compactness, is preferable to long, open wool, since it constitutes a better safeguard from inclemencies of weather, and better protects the animal from the bad effects of cold and drenching rains in spring and fall. The wool should be, as nearly as possible, of even length and thickness over the entire body. Shortness on the flank, and shortness or thickness on the belly, are serious defects.

Evenness of fleece is a point of the first importance. Many sheep exhibit good wool on the shoulder and side, while it is far coarser and even hairy on the thighs, dew-lap, etc. Rams of this stamp should not be bred from by any one aiming to establish a superior fine-woolled flock; and all such ewes should gradually be excluded from those selected for breeding.

The style of the wool is a point of as much importance as mere fineness. Some very fine wool is stiff, and the fibres almost straight, like hair. It has a dry, cottony look; and is a poor, unsalable article, however fine the fibre. Softness of wool—a delicate, silky, highly elastic feel between the fingers or on the lips, is the first thing to be regarded. This is usually an index, or inseparable attendant, of the other good qualities; so much so, indeed, that an experienced judge can decide, with little difficulty, between the quality of two fleeces, in the dark. Wool should be finely serrated, or crimped from one extremity to the other: that is, it should present a regular series of minute curves; and, generally, the greater the number of these curves in a given length, the higher the quality of the wool in all other particulars. The wool should open on the back of the sheep in connected masses, instead of breaking up into little round spiral ringlets of the size of a pipe-stem, which indicate thinness of fleece; and when the wool is pressed open each way with the hands, it should be close enough to conceal all but a delicate rose-colored line of skin. The interior of the wool should be a pure, glittering white, with a lustre and liveliness of appearance not surpassed in the best silk.

The points in the form of the Merino which the breeder is called upon particularly to avoid are, a long, thin head, narrow between the eyes; a thin, long neck, arching downward before the shoulders; narrow loins; flat ribs; steep, narrow hind-quarters; long legs; thighs scarcely meeting at all; and legs drawn far under the body at the least approach of cold. All these points were, separately or conjointly, illustrated in many of the Saxon flocks which have been swept from the country. Sufficient attention has already been paid to the points to be avoided in the fleece.

BREEDING MERINOS.

The first great starting-point, among pure-blood animals, is, that "like will beget like." If the sire and ewe are perfect in any given points, the offspring will generally be; if either is defective, the offspring—subject to a law which will possibly be noticed—will be half-way between the two; if both are defective in the same points, the progeny will be more so than either of its parents—it will inherit the amount of defect in both parents added together. There are exceedingly few perfect animals. Breeding, therefore, is a system of counter-balancing—breeding out—in the offspring, the defects of one parent, by the marked excellence of the other parent, in the same points. The highest blood confers on the parent possessing it the greatest power of stamping its own characteristics on its progeny; but, blood being the same, the male sheep possesses this power in a greater degree than the female. We may, therefore, in the beginning, breed from ewes possessing any defects short of cardinal ones, without impropriety, provided we possess the proper ram for that purpose; but, where a high standard of quality is aimed at, all ewes possessing even considerable defects should gradually be thrown out from breeding. Every year should add to the vigor of the selection.

But, from the beginning—and at the beginning, more than at any other time—the greatest care should be evinced in the selection of the ram. If he has a defect, that defect is to be inherited by the whole future flock; if it is a material one—as, for example, a hollow back, bad cross, or thin fleece, or a highly uneven fleece—the flock will be one of low quality and little value. If, on the other hand, he is perfect, the defects in the female will be lessened, and gradually bred out. It being, however, difficult to find perfect rams, those should be taken which have the fewest and lightest defects, and none of these material, like those just enumerated. These defects are to be met and counterbalanced by the decided excellence—sometimes, indeed, running into a fault—of the ewe, in the same points. If the ram, then, is a little too long-legged, the shortest-legged ewes should be selected for him; if gummy, the dryest-woolled; if his fleece is a trifle below the proper standard of fineness—but he has been retained, as often happens, for weight of fleece and general excellence—he is to be put to the finest and lightest-fleeced ewes, and so on. With a selection of rams, this system of counterbalancing would require but little skill, if each parent possessed only a single fault. If the ewe be a trifle too thin-fleeced, and good in all other particulars, it would require no nice judgment to decide that she should be bred to an uncommonly thick-fleeced ram. But most animals possess, to a greater or less degree, several defects. To select so that every one of these in the dam shall meet its opposite in the male, and the contrary, requires not only plentiful materials from which to select, but the keenest discrimination.

After the breeder has successfully established his flock, and given them an excellent character, he soon encounters a serious evil. He must "breed in-and-in," as it is called—that is, interbreed between animals more or less nearly related in blood—or he must seek rams from other flocks, at the risk of losing or changing the distinctive character of his flock, hitherto so carefully sought, and built up with so much painstaking. The opponents of in-and-in breeding contend that it renders diseases and all other defects hereditary, and that it tends to decrease of size, debility, aud a general breaking up of the constitution. Its defenders, on the other hand, insist that, if the parents are perfectly healthy, this mode does not, of itself, tend to any diminution of healthfulness in the offspring; and they likewise claim—which must be conceded

—that it enables the skilful breeder much more rapidly to bring his flock to a particular standard or model, and to keep it there much more easily—unless it be true that, in course of time, they will dwindle and grow feeble.

So far as the effect on the constitution is concerned, both positions may be, to a certain extent, true. But it is, perhaps, difficult always to decide with certainty when an animal is not only free from disease, but from all tendency or predisposition towards it. A brother or sister may be apparently healthy—may be actually so—but may still possess a peculiarity of individual conformation which, under certain circumstances, will manifest itself. If these circumstances do not chance to occur, they may live until old age, apparently possessing a robust constitution. If tried together, their offspring—by a rule already laid down—will possess this individual tendency in a double degree. If the ram be interbred with sisters, half-sisters, daughters, granddaughters, etc., for several generations, the predisposition toward a particular disease—in the first place slight, now strong, and constantly growing stronger—will pervade, and become radi-

THE SCOTCH SHEEP-DOG, OR COLLEY.

cally incorporated into, the constitution of the whole flock. The first time the requisite exciting causes are brought to bear, the disease breaks out, and, under such circumstances, with peculiar severity and malignancy. If it be of a fatal character, the flock is rapidly swept away; if not, it becomes chronic, or periodical at frequently recurring intervals. The same remarks apply, in part, to those defects of the outward form which do not at first, from their slightness, attract the notice of the ordinary breeder. They are rapidly increased until, almost before thought of by the owner, they destroy the value of the sheep. That such are the common effects of in-and-in breeding, with such skill as it is commonly conducted, all know who have given attention to the subject; and for these reasons the system is regarded with decided disapprobation and repugnance by nine out of ten of the best practical farmers.

The sheep-breeder can, however, avoid the effects of in-and-in breeding, and at the same time preserve the character of his flock, by seeking rams of the same breed, possessing, as nearly as possible, *the characteristics which he wishes to preserve in his own flock.* If this rule is neglected—if he draws indiscriminately from all the different varieties or families of a breed—some large, and some small—some long-woolled, and some short-woolled—some medium, and some superfine in quality—some tall, and some squatty—some crusted over with black gum, and some entirely free from it—breeding will become a mere matter of hap-hazard, and no certain or uniform results can be expected. So many varieties cannot be fused into one for a number of generations—as is evidenced by the want of uniformity in the Rambouillet flock, which was commenced by a promiscuous admixture of all the Spanish families; and it not merely happens, as between certain classes of Saxons, that particular families can never be successfully amalgamated.

If, however, the breeder has reached no satisfactory standard—if his sheep are deficient in the requisites which he desires—he is still to adhere to the breed—*provided the desired requisites are characteristic of the breed he possesses*—and select better animals to improve his own inferior ones. If he has, for instance, an inferior flock of South-Downs, and wishes to obtain the qualities of the best Down dams, he should seek for the best rams of that breed. But if he wishes to obtain qualities *not characteristic of the breed he possesses,* he must cross with a breed which does possess them. If the possessor of South-Downs wishes to convert them into a fine-woolled sheep similar to the Merino, he should cross his flock steadily with Merino rams—constantly increasing the amount of Merino, and diminishing the amount of South-Down blood. To effect the same result, he would take the same course with the common sheep of the country, or with any other coarse race.

There are those, who, forgetting that some of the finest varieties now in existence, of several kinds of domestic animals, are the result of crosses—bitterly inveigh against the practice of crossing, under any and all circumstances. It is, it must be admitted, an unqualified absurdity, as frequently conducted—as, for example, an attempt to unite the fleece of a Merino and the carcass of a Leicester, by crosses between those breeds; but, under the limitations already laid down, and with the objects specified as legitimate ones, this objection to crossing savors of the most profound prejudice, or the most unblushing quackery. It is neither convenient, nor within the means of every man wishing to start a flock of sheep, to commence exclusively with full-bloods. With a few to breed rams from, and to begin a full-blood stock, the breeder will find it his best policy to purchase the best common sheep of his country, and gradually grade them up with Merino rams. In selecting the ewes, good shape, fair size, and a robust constitution, are the main points—the little difference in the quality of the common sheep's wool being of no consequence. For their wool, they are to look to the Merino; but good form and constitution they can and ought to possess, so as not to entail deep-rooted and entirely unnecessary evils on their progeny.

Satisfactory results have followed crossing a Down ram—small, compact, exceedingly beautiful, fine and even fleeced—with large-sized Merino ewes. The half-blood ewes were then bred to a Merino ram, and also their female progeny, and so on. The South-Downs, from a disposition to take on fat, manifested themselves, to a perceptible extent, in every generation, and the wool of many of the sheep in the third generation—seven-eighths blood Merino, and one-eighth blood Down—was very even, and equal to medium, and some of them to good medium Merino. Their fleeces were lighter than the full-blood Merinos, but increased in weight with each succeeding cross back toward the latter. The mutton of the first, and even of the second cross was of a beautiful flavor, and retained, to the last, some of the superiority of South-Down mutton.

Results are also noted of breeding Leicester ewes—taking

one cross of the blood, as in the preceding case—toward the Merino. The mongrels, to the second generation—beyond which they were not bred—were about midway between the parent stock in size—with wool shorter, but far more fine and compact than the Leicester—their fleeces about the same weight, five pounds—and, altogether, they were a showy and profitable sheep, and well calculated to please the mass of farmers. Their fleeces, however, lacked evenness, their thighs remaining disproportionately coarser and heavy.

A difference of opinion exists in relation to the number of crosses necessary before it is proper to breed from a mongrel ram. Some high authorities assert that it does not admit of the slightest doubt that a Merino, in the fourth generation, from even the worst-woolled ones, is in every respect equal to the stock of the sire—that no difference need to be made in the choice of a ram, whether he is a full-blood, or a fifteen-sixteenths—and that, however coarse the fleece of the parent ewe may have been, the progeny in the fourth generation will not show it.

Others, however—while admitting that the only value of blood or pedigree, in breeding, is to insure the hereditary transmission of the properties of the parent to the offspring, and that, as soon as a mongrel reaches the point where he stamps his characteristics on the progeny, with the same certainty that a full-blood does, he is equally valuable, provided he is, individually, as perfect an animal—contend that this cannot be depended upon, with any certainty, in rams of the fourth Merino cross. They assert that the offspring of such crosses invariably lack the style and perfection of thorough-bred flocks. The sixth, seventh, or eighth cross might be generally, and the last, perhaps, almost invariably, as good as pure-blood rams; yet pure blood is a fixed standard, and were every breeder to think himself at liberty to depart from it in his rams, each one more or less, according to his judgment or caprice, the whole blood of the country would become adulterated. No man, assuredly, can be authorized to sell a ram of any cross, whether the tenth, or even the twentieth, as a full-blood.

It is of the utmost importance for those *commencing* flocks, either of full-bloods, or by crossing, to select the choicest rams. A grown ram may, by methods which will hereafter be described, be made to serve from one hundred to one hundred and fifty ewes in a season. A good Merino ram will, moderately speaking, add more than a pound of wool to the fleece of the dam, or every lamb got by it, from a common-woolled ewe—that is, if the ewe at three years old sheared three pounds of wool, the lamb at the same age will shear four. This would give one hundred or one hundred and fifty pounds of wool for the use of a ram for a single season; and every lamb subsequently got by him adds a pound to this amount. Many a ram gets, during his life, eight hundred or one thousand lambs. Nor is the extra amount of wool all. He gets from eight hundred to one thousand half-blooded sheep, worth double their dams, and ready to be made the basis of another and higher stride in improvement. A good ram, then, is as important and, it may be, quite as valuable an animal as a good farm-horse stallion. When the number of a ram's progeny are taken into consideration, and when it is seen over what an immense extent, even in his own direct offspring, his good or bad qualities are to be perpetuated, the folly of that economy which would select an inferior animal is sufficiently obvious.

It will be found the best economy in starting a flock, where the proper flocks from which to draw rams are not convenient, to purchase several of the same breed, of course, but *of different strains of blood.* Thus ram No. 2 can be put on the offspring of No. 1, and the reverse; No. 3 can be put on the offspring of both, and both upon the offspring of No. 3. The changes which can be rung on three distinct strains of blood, without in-and-in breeding close enough to be attended with any considerable danger, are innumerable.

The brother and sister, it will be born in mind, are of the same blood; the father and daughter, half; the father and granddaughter, one-fourth; the father and great-granddaughter, one-eighth; and so on. Breeding between animals possessing one-eighth of the same blood, would not be considered very close breeding; and it is not unusual, in rugged, well-formed families, to breed between those possessing one-fourth of the same blood.

If, however, these rams of different strains are brought promiscuously, without reference to similarity of characteristics, there may, and probably will, be difference between them; and it might require time and skill to give a flock descended from them a proper uniformity of character. Those who breed rams for sale should be prepared to furnish different strains of blood, with the necessary individual and family uniformity.

GENERAL PRINCIPLES OF BREEDING.

Some few suggestions upon the general principles to be observed in breeding may not be superfluous here, referring the reader, who is disposed to investigate this subject in detail, to its full discussion in the author's treatise upon "Cattle and their Diseases."

As illustrative of the importance of *breeding only from the best*, taking care to avoid structural defects, and especially to secure freedom from *hereditary diseases*, since both defects and diseases appear to be more easily transmissible than desirable qualities, it may be remarked that scrofula is not uncommon among sheep, and presents itself in various forms. Sometimes it is connected with consumption; sometimes it affects the viscera of the abdomen, and particularly the mesenteric glands, in a manner similar to consumption in the lungs. The scrofulous taint has been known to be so strong as to affect the fœtus, and lambs have occasionally been dropped with it; but much oftener they show it at an early age, and any affected in this way are liable to fall an easy prey to any ordinary or prevalent disease, which develops in such with unusual severity. Sheep are also liable to several diseases of the brain, and of the respiratory and digestive organs. Epilepsy, or "fits," and rheumatism sometimes occur.

The breeder's aim should be to grasp and *render permanent*, and increase so far as practicable, *every variation for the better*, and to reject for breeding purposes such as show a downward tendency. A remarkable instance of the success which has often attended the well-directed efforts of intelligent breeders, is furnished in the new Mauchamp-Merino sheep, which originated in a single animal—a product of the law of variation—and which, by skilful breeding and selection, has become an established breed of a peculiar type, and possessing valuable properties. Samples of the wool of these sheep were shown at the great exhibition in London, in 1851, as well as at the subsequent great agricultural exhibition at Paris, and attracted much attention.

This breed was originated by Mons. J. L. Graux. In 1828, a Merino ewe produced a peculiar ram lamb, having a different shape from the ordinary Merino, and possessing wool singularly long, straight, and silky. Two years afterward, Mr. Graux obtained by this ram one ram and one ewe, having the silky character of wool. Among the produce of the ensuing year were four rams and one ewe with similar fleeces; and in 1833, there were rams enough of the new sort to serve the whole flock of ewes. In each subsequent year, the lambs were of two kinds; one possessing the curled, elastic wool of the old Merinos, only a little longer and finer, and the other like the new breed. At last, the skilful breeder obtained a flock containing the fine, silky fleece with a smaller breed, broader flanks, and more capacious chest; and several flocks being crossed with the Mauchamp variety, the Mauchamp-Merino breed is the result.

The pure Mauchamp wool is remarkable for its qualities as a combing-wool, owing to the strength, as well as the length and fineness of the fibre. It is found of great value by the manufacturers of Cashmere shawls, and similar goods, being second only to the true Cashmere fleece, in the fine, flexible delicacy of the fibre; and when in combination with Cashmere wool, imparting strength and consistency. The quantity of this wool has since become as great as that from ordinary Merinos, or greater, while its quality commands twenty-five per cent. higher price in the French market. Breeders, certainly, cannot watch too closely any accidental peculiarity of conformation or characteristic in their flocks.

The apparent influence of the male first having fruitful intercourse with a female, *upon her subsequent offspring by other males*, has been noticed by various writers. The following well-authenticated instances are in point:

A small flock of ewes, belonging to Dr. W. Wells, in the island of Granada, was served by a ram procured for the purpose. The ewes were all white and woolly; the ram was quite different, being of a chocolate color, and hairy like a goat. The progeny were, of course, crosses, but bore a strong resemblance to the male parent. The next season, Dr. Wells obtained a ram of precisely the same breed as the ewes; but the progeny showed distinct marks of resemblance to the former ram, in color and covering. The same thing occurred on neighboring estates, under like circumstances.

Six very superior pure-bred black-faced horned ewes, belonging to Mr. H. Shaw, of Leochel, Cushnie, were served by a white-faced hornless Leicester ram. The lambs were crosses. The next year they were served by a ram of exactly the same breed as the ewes themselves, and their lambs were, without an exception, hornless and brownish in the face, instead of being black and horned. The third year they were again served by a superior ram of their own breed; and again the lambs were mongrels, but showed less of the Leicester characteristics than before; and Mr. Shaw at last parted from these fine ewes without obtaining a single pure-bred lamb.

To account for this result—seemingly regarded by most physiologists as inexplicable—Mr. James McGillivray, V. S., of Huntley, has offered an explanation, which has received the sanction of a number of competent writers. His theory is, that when a pure animal of any breed has been pregnant by an animal of a different breed, such pregnant animal *is a cross ever after*, the purity of her blood being lost, in consequence of her connection with the foreign animal, and herself becoming a cross forever, incapable of producing a pure calf of any breed.

To cross, *merely for the sake of crossing*, to do so without that care and vigilance which are highly essential, is a practice which cannot be too much condemned, being, in fact, a national evil, if pushed to such an extent as to do away with a useful breed of animals, and establish a generation of mongrels in their place—a result which has followed in numerous instances amongst every breed of animals.

The principal use of crossing is to raise animals for the butcher. The male, being generally an animal of a superior breed, and of a vigorous nature, almost invariably stamps his external form, size, and muscular development on the offspring, which thus bear a strong resemblance to him; while their internal nature, derived from the dam, well adapts them to the locality, as well as to the treatment to which their dams have been accustomed.

With sheep, where the peculiarities of the soil, as regards the goodness of feed, and exposure to the severities of the weather, often prevent the introduction of an improved breed, the value of using a new and superior ram is often very considerable; and the weight of mutton is thereby materially increased, without its quality being impaired, while earlier maturity is at the same time obtained. It involves, however, more systematic attention than most farmers usually like to bestow, for it is necessary to employ a different ram for each purpose; that is, a native ram, for a portion of the ewes to keep up the purity of the breed, and a foreign ram, to raise the improved cross-bred animals for felting, either as lambs or sheep. This plan is adopted by many breeders of Leicester sheep, who thus employ South-Down rams to improve the quality of the mutton.

One inconvenience attending this plan is the necessity of fattening the maiden ewes as well as the wethers. They may, however, be disposed of as fat lambs, or the practice of spaying (fully explained in "Cattle and their Diseases") might be adopted, so as to increase the felting disposition of the animal. Crossing, therefore, should be adopted with the greatest caution and skill, where the object is to improve the breed of animals. It should never be practised carelessly or capriciously, but it may be advantageously pursued, with a view to raising superior and profitable animals for the butcher. For the latter purpose, it is generally advisable to use males of a larger breed, provided they possess a disposition to fatten; yet, in such cases, it is of importance that the *pelvis* of the female should be wide and capacious, so that no injury may arise in lambing, in consequence of the increased size of the heads of the lambs. The shape of the ram's head should be studied, for the same reason.

In crossing, however, for the purpose of establishing a new breed, the size of the male must give way to other more important considerations; although it will still be desirable to use a large female of the breed which is sought to be improved. Thus, the South-Downs have vastly improved the larger Hampshires, and the Leicester, the huge Lincolns and the Cotswolds.

USE OF RAMS.

Merino rams are frequently used from the first to the tenth year, and even longer. The lambs of very old rams are commonly supposed not to be as those of middle-aged ones; though where rams have not been overtasked, and have been properly fed, little if any difference is discoverable in their progeny by reason of their sire's age. A ram lamb should not be used, as it retards his growth, injures his form, and, in many instances, permanently impairs his vigor and courage A yearling may run with thirty ewes, a two-year-old with from forty to fifty, and a three-year-old with from fifty to sixty; while some very powerful, mature rams will serve seventy or eighty. Fifty, however, is enough, where they *run with* the ewes. It is well settled that an impoverished and overtasked animal does not transmit his individual properties so decidedly to his offspring as does one in full vigor.

Rams, of course, are not to be selected for ewes by mere chance, but according as their qualities may improve those of the ewes. It may not be superfluous, though seemingly a repetition, to state that a good ewe flock should exhibit these characteristics: *strong bone*, supporting a roomy frame, affording space for a large development of flesh; *abundance of wool of a good quality*, keeping the ewes warm in inclement weather, and insuring profit to the breeder; *a disposition to fatten early*, enabling the breeder readily to get rid of his sheep selected for the butcher; and *a prolific tendency*, increasing the flock rapidly, and being also a source of profit. Every one of these properties is advantageous in itself; but when all are combined in the same individuals of a flock, that flock is in a high state of perfection. In selecting rams, it should be observed whether or not they possess one or more of those qualities in which the ewes may be deficient, in which case their union with the ewes will produce in the progeny a higher degree of perfection than is to be found in the ewes themselves, and such a result will improve the state of the future ewe-flock; but, on the contrary, if the ewes are superior in all points to the rams, then, of course, the use of such will only serve to deteriorate the future ewe-flock.

Several rams running in the same flock excite each other to an unnatural and unnecessary activity, besides injuring each other by constant blows. It is, in every point of view, bad husbandry, where it can be avoided, and, as customarily managed, is destructive to every thing like careful and judicious breeding. The nice adaptation which the male should possess to the female is out of the question where half a dozen or more rams are running promiscuously with two or three hundred ewes.

Before the rams are let out, the breeding ewes should all be brought together in one yard; the form of each noted, together with the length, thickness, quality and style of her wool—ascertained by opening the wool on the shoulder, thigh, and belly. When every point is thus determined, that ram should be selected which, on the whole, is best calculated to perpetuate the excellencies of each, both of fleece and carcass, and to best counterbalance defects in the mutual offspring. Every ewe, when turned in with the ram, should be given a distinct mark, which will continue visible until the next shearing. For this purpose, nothing is better than Venetian red and hog's lard, well incorporated, and marked on with a cob. The ewes for each ram require a differently shaped mark, and the mark should also be made on the ram, as noted in the sheep-book. Thus it can be determined at a glance by what ram the ewe was tapped, any time before the next shearing. The ewes selected for each ram are placed in different enclosures, and the chosen ram placed with them. Rams require but little preparation on being put among ewes. If their skin is red in the flanks when the sheep are turned up, they are ready for the ewes, for the natural desire is then upon them. Most of the ewes will be served during the second week the ram is among them, and in the third, all. It is better, however, not to withdraw the rams until the expiration of four weeks, when the flocks can be doubled, or otherwise re-arranged for winter, as may be necessary. The trouble thus taken is, in reality, slight—nothing, indeed, when the beneficial results are considered. With two assistants, several hundred ewes may be properly classified and divided in a single day.

Where choice rams are scarce, so that it is desirable to make the services of one go a great way, or where it is impossible to have separate enclosures—as on farms where there are a great number of breeding ewes, or where the shepherd system is adopted, to the exclusion of fences—resort may be had to another method. A hut should be built, containing as many apartments as the ram is desired to be used, with an alley between them, each apartment to be furnished with a feeding-box and trough in one corner, and gates or bars opening from each into the alley, and at each end of the alley. Adjoining these apartments, a yard should be inclosed, of size just sufficient to hold the flock of breeding ewes.

A couple of strong rams, of any quality, for about every

hundred ewes, are then aproned, their briskets rubbed with Venetian red and hog's lard, and let loose among the ewes. *Aproning* is performed by sewing a belt of coarse sacking, broad enough to extend from the fore to the hind legs, loosely but strongly around the body. To prevent its slipping forward or back, straps are carried round the breast and back of the breech. It should be made *perfectly secure*, or all the labor of this method of coupling will be far worse than thrown away. The pigment on the brisket should be renewed every two or three days; and it will be necessary to change the "teasers"—as these aproned rams are called—about once a week, as they do not long retain their courage under such unnatural circumstances. Twice a day the ewes are brought into the yard in front of the hut. Those marked on their rumps by the teasers are taken into the alley. Each is admitted *once* to the ram for which she is marked, and then goes out *at the opposite end of the alley* from which they entered, into a field separate from that containing the flock from which she was taken. A powerful and vigorous ram, from three to seven years old, and properly fed, can thus be made to serve from one hundred and fifty to even two hundred ewes, with no greater injury than from running loose with fifty or sixty. The labor here required is likewise more apparent than real, when the operation is conducted in a systematic manner.

Rams will do better, accomplish more, and last two or three years longer, if daily fed with grain, when on service, and it is better to continue it. In all cases, they should, after serving, be put on good pasture, as they will have lost a good deal of condition, being indisposed to settle during the tapping season.

A ram should receive the equivalent of from half a pint to a pint of oats daily, when worked hard. They are much more conveniently fed when kept in huts. If suffered to run at large, they should be so thoroughly tamed that they will eat from a measure held by the shepherd. Careful breeders thus train their stock-rams, from the time they are lambs. It is very convenient, also, to have them halter-broke, so that they can be led about without dragging or lifting them. An iron ring attached to one of the horns, near the point, to which a cord can be fastened for leading, confining, etc., is very useful and convenient. If rams are wild, it is a matter of considerable difficulty to feed them separately, and it can only be effected by yarding the flock and catching them out. Some breeders, in addition to extra feeding, take the rams out of the flocks each night, shutting them up in a barn or stable by themselves. To this practice there is no objection, and it greatly saves their strength.

Rams should not be suffered to run with the ewes over a month, at least in the Northern States. It is much better that a ewe go dry than that she have a lamb later than the first of June. Besides, after the rutting season is over, the rams grow cross, frequently striking the pregnant ewes dangerous blows with their heavy horns, at the racks and troughs.

It is reasonably enough conjectured, that if procreation and the first period of gestation take place in cold weather, the fœtus will be fitted for the climate which rules during the early stages of its existence. If this be so—and it is certainly in accordance with the laws of Nature—fine-woolled sheep are most likely to maintain their excellence by deferring the connection of the male till the commencement of cold weather;

and, in the Northern States, this is done about the first of December, thus bringing the yeaning time in the last of April, or the first of May, when the early grass affords a large supply and good quality of food.

LAMBING.

The ewe goes with young about five months, varying from one hundred and forty-five to one hundred and sixty-two days. Pregnant ewes require the same food as at all other times. Until two or three weeks preceding lambing, it is only necessary that they, like other store-sheep,

EWE AND LAMBS.

be kept in good, plump, ordinary condition; nor are any separate arrangements necessary for them after that period, in a climate where they obtain sufficient succulent food to provide for a proper secretion of milk. In backward seasons in the North, where the grass does not start prior to the lambing-time, careful farmers feed their ewes on chopped roots, or roots mixed with oat and pea-meal, which is excellent economy. Caution is, however, necessary to prevent injury or abortion, which is often the result of excessive fat, feebleness, or disease. The first may be remedied by blood-letting and spare diet; and both the last by restored health and generous food. Sudden frights, as from dogs or strange objects; long or severe journeys, great exertions, unwholesome food, blows in the region of the fœtus, and some other causes, produce abortion.

Lambs are usually dropped, in the North, from the first to the fifteenth of May; in the South, they can safely come earlier. It is not expedient to have them dropped when the weather is cold or boisterous, as they require too much care; but the sooner the better, after the weather has become mild, and the herbage has started sufficiently to give the ewes that green food which is required to produce a plentiful secretion of milk. It is customary, in the North, to have fields of clover, or the earliest grasses, reserved for the early spring-feed of the breeding-ewes; and, if these can be contiguous to their stables, it is a great convenience—for the ewes should be confined in the latter, on cold and stormy nights, during the lambing season.

If the weather be warm and pleasant, and the nights moderately warm, it is better to have the lambing take place in the pasture; since sheep are then more disposed to own their lambs, and take kindly to them, than in the confusion of a small inclosure. In the latter, sheep, unless particularly docile, crowd from one side to another when any one enters, running over young lambs, pressing them severely, etc.; ewes become separated from their lambs, and then run violently round from one to another, jostling and knocking them about; young and timid ewes, when so separated, will frequently neglect their lambs for an hour or more before they will again approach them, while, if the weather is severely cold, the lamb, it it has never sucked, is in danger of perishing. Lambs, too, when first dropped in a *dirty* inclosure, tumble about, in their first efforts to rise, and the membrane which adheres to them becomes smeared with dirt and dung; and the ewe's refusing to lick them dry much increases the hazard of freezing

In cold storms, however, and in sudden and severe weather, all this must be encountered; and, therefore, every shepherd should teach his sheep docility. It requires but a very moderately cold night to destroy the new-born Saxon lamb, which—the pure blood—is dropped nearly as naked as a child. During a severely cold period, of several days continuance, it is almost impossible to rear them, even in the best shelter. The Merino, South-Down, and some other breeds, will endure a greater degree of cold with impunity. Where inclosures are used for yeaning, they should be kept clean by frequent litterings of straw—not enough, however, to be thrown on at any one time, to embarrass the lamb about rising.

The predisposing symptoms of lambing are, enlargement and reddening of the parts under the tail, and drooping of the flanks. The more immediate are, when the ewe stretches herself frequently; separating herself from her companions; exhibiting restlessness by not remaining in one place for any length of time; lying down and rising up again, as if dissatisfied with the place; pawing the ground with a forefoot; bleating, as if in quest of a lamb; and appearing fond of the lambs of other ewes. In a very few hours, or even shorter time, after the exhibition of these symptoms, the immediate symptom of lambing is the expulsion of the bag of water from the *vagina*. When this is observed, the ewe should be narrowly watched, for the pains of labor may be expected to come on immediately. When these are felt by her, the ewe presses or forces with earnestness, changing one place or position for another, as if desirous of relief.

The ewe does not often require mechanical assistance in parturition. Her labors will sometimes be prolonged for three or four hours, and her loud moanings will evince the extent of her pain. Sometimes she will go about several hours, and even resume her grazing, with the fore-feet and nose of the lamb protruding at the mouth of the *vagina*. If let alone, however, Nature will generally relieve her. In case of a false parturition of the fœtus—which is comparatively rare—the shepherd may apply his thumb and finger, after oiling, to push back the lamb, and assist in gently turning it till the nose and forefeet appear. Where feebleness in expelling the fœtus exists, only the slightest aid should be rendered, and that to help the throes of the dam. The objection to interfering—except as a last resort—is, that the ewe is frightened when caught, and her efforts to expel the lamb cease. When aided, in any case, the gentlest force should be applied, and only in conjunction with the efforts of the ewe. The clearing, or *placenta*, generally drops from the ewe in the course of a very short time—in many cases, within a few minutes—after lambing. It should be carried away, and not allowed to lie upon the lambing-pound.

Common kale, or curly-greens, is excellent food for ewes that have lambed, as its nutritive matter, being mucilaginous, is wholly soluble in water, and beneficial in encouraging the necessary discharges of the ewe at the time of lambing. In these respects, it is a better food than Swedish turnips—upon which sheep are sometimes fed—which become rather too fibrous and astringent, in spring, for the secretion of milk. In the absence of kale or cabbage, a little oil-cake will aid the discharges and purify the body. New grass also operates medicinally upon the system.

MANAGEMENT OF LAMBS.

While the lamb is tumbling about and attempting to rise—the ewe, meanwhile, licking it dry—it is well to be in no haste to interfere. A lamb that gets at the teat without help, and procures even a small quantity of milk, knows how to help itself afterward, and rarely perishes. If helped, it sometimes continues to expect it, and will do little for itself for two or three days. The same is true where lambs are fed from a spoon or bottle.

But if the lamb ceases to make efforts to rise—especially if the ewe has left off licking it while it is wet and chilly—it is time to render assistance. It is not advisable to throw the ewe down—as is frequently practised—in order to suckle the lamb; because instinct teaches the latter to point its nose *upward* in search of the teats. It is, therefore, doubly difficult to teach it to suck from the bag of the prostrate ewe; and when it is taught to do this, by being so suckled several times, it is awkward about finding the teat in the natural position, when it begins to stand and help itself. Carefully disengaging the ewe from her companions, with his crook—which useful article will be hereafter described—the assistant should place one hand before the neck and the other behind the buttocks of the ewe, and then, pressing her against his knees, he should hold her firmly and still, so that she will not be constantly crowding away from the shepherd, who should set the lamb on its feet, inducing it to stand, if possible; if not, supporting it *on its feet* by placing one hand under its body; put its mouth to the teat, and encourage it to suck by tickling it about the roots of the tail, flanks, etc., with a finger. The lamb, mistaking this last for the caresses of its dam, will redouble its efforts to suck. Sometimes it will manifest great dullness, and even apparent obstinacy, in refusing for a long time to attempt to assist itself, crowding backward, etc.; but the kind and gentle shepherd, who will not sink himself to the level of a brute, by resenting the stupidity of a brute, will generally carry the point by perseverance. Sometimes milking a little into the lamb's mouth, holding the latter close to the teat, will induce it to take hold.

If the ewe has no milk, the lamb should be fed, until the natural supply commences, with small quantities of the milk of a *new-milch* cow. This should be mixed, say half and half, with water, with enough molasses to give it the purgative effect of the first milk, gently warmed to the natural heat—not scalded and suffered to cool—and then fed through a bottle with a sponge in the opening of it, which the lamb should *suck*, if it can be induced so to do. If the milk is poured in its mouth from a spoon or bottle, it is frequently difficult, as before stated, to induce it to suck. Moreover, unless milk is poured into the mouth slowly and with care—no faster than the lamb can swallow—a speedy wheezing, the infallible precursor of death, will show that a portion of the fluid has been forced into the lungs. Lambs have been frequently killed in this way.

If a lamb becomes chilled, it should be wrapped in a woollen blanket, placed in a warm room, and given a little milk as soon as it will swallow. A trifle of pepper is sometimes placed in the milk, and with good effect, for the purpose of rousing the cold and torpid stomach into action. In New England, under such circumstances, the lamb is sometimes "baked," as it is called—that is, put in a blanket in a moderately-heated oven, until warmth and animation are restored; others immerse it in tepid water, and subsequently rub it dry, which is said to be an excellent method where the lamb is nearly frozen. A good blanket however, a warm room, and sometimes, perhaps, a little gentle friction will generally suffice.

If a strong ewe, with a good bag of milk, chance to lose her lamb, she should be required to bring up one of some other ewe's twins, or the lamb of some feeble or young ewe, having an inadequate supply of milk. Her own lamb should be skinned as soon as possible after death, and the skin sewed over the lamb which she is to foster. She will sometimes be a little suspicious for a day or two; and if so, she should be kept in a small pen with the lamb, and occasionally looked to. After she has taken well to it, the false skin may be removed in three or four days. If no lamb is placed on a ewe which lost her lamb, and which has a full bag of milk, the milk should be drawn from the bag once or twice, or garget may ensue; even if this is not the result, permanent indurations, or other results of inflammatory action, will take place, injuring the subsequent nursing properties of the animal.

When milked, it is well to wash the bag for some time in cold water, since it checks the subsequent secretions of milk, as well as allays inflammation.

Sometimes a young ewe, though exhibiting sufficient fondness for her lamb, will not stand for it to suck; and in this case, if the lamb is not very strong and persevering, and particularly if the weather is cold, it soon grows weak, and perishes. The conduct of the dam, in such instances, is occasioned by inflammatory action about the bag or teats, and perhaps somewhat by the novelty of her position. In this case, the sheep should be caught and held until the lamb has exhausted her bag, and there will not often be any trouble afterward; though it may be well enough to keep them in a pen together until the fact is determined.

Such pens—necessary in a variety of cases other than those mentioned—need not exceed eight or ten feet square, and should be built of light materials, and fastened together at the corners, so that they can be readily moved by one man, or, at the most, two, from place to place, where they are wanted. Their position should be daily shifted, when sheep are in them, for cleanliness and fresh feed. Light pine poles, laid up like a fence, and each nailed and pegged to the lower ones at the corners, or laid on, are quite serviceable. Two or three sides of a few of them should be wattled with twigs, and the tops partly covered, in order to shield feeble lambs from cold rains, piercing winds, and the like.

Young lambs are subject to what is commonly known as "pinning"—that is, their first excrements are so adhesive and tenacious that the orifice of the anus is closed, and subsequent evacuations prevented. The adhering matter, in such cases, should be entirely removed, and the part rubbed with a little dry clay, to prevent subsequent adhesion. Lambs will frequently perish from this cause, if not looked to for the first few days.

The ewes and their young ought to be divided into small flocks, and have a frequent change of pasture. Some careful shepherds adopt the plan of confining their lambs, allowing them to suck two or three times a day. By this method they suffer no fatigue, and thrive much faster. It is, however, troublesome as well as injurious, since the exercise is essential to the health and constitution of the lamb intended for rearing. It is admissible only when they are wanted for an early market; and with those who rear them for this purpose it is a common practice.

Where there are orphans or supernumeraries in the flock, the deserted lambs must be brought up by hand. Such animals, called pet lambs, are supported on cow's milk, which they receive warm from the cows each time they are milked, and as much as they can drink. In the intervals of meals, in bad weather, they are kept under cover; in good weather they are put into a grass enclosure during the day, and sheltered at night until the nights become warm. They are fed by hand out of a small vessel, which should contain as much milk as it is known each can drink. They are first taught to drink out of the vessel with the fingers, like a calf, and as soon as they can hold a finger steady in the mouth, a small tin tube, about three inches in length, and of the thickness of a goose-quill, should be covered with several folds of linen, sewed tightly on, to use as a substitute for a teat, by means of which they will drink their allowance of milk with great ease and quickness. A goose-quill would answer the same purpose, were it not easily squeezed together by the mouth. When the same person feeds the lambs—and this should be the dairy-maid—they soon become attached to her, and desire to follow her everywhere; but to prevent their bleating, and to make them contented, an apron or a piece of cloth, hung on a stake or bush in the inclosure, will keep them together.

It is much better for the lambs and for their dams that they be *weaned* from three and a half to four months old. When taken away, they should be put for several days in a field distant from the ewes, that they may not hear each other's bleatings, as the lambs, when in hearing of their dams, continue restless much longer, and make constant and, frequently, successful efforts to crawl through the fences which separate them. One or two tame old ewes are turned into the field with them, to teach them to come at the call, find salt when thrown to them, and eat out of troughs when winter approaches.

When weaned, the lambs should be put on the freshest and tenderest grass—rich, sweet food, but not too luxuriant. The grass and clover, sown the preceding spring, on grain-fields seeded down, is often reserved for them. The dams, on the contrary, should be put for a fortnight on short, dry feed, to stop the flow of milk. They should be looked to after a day or two, and if the bags of any are found much distended, the milk should be drawn away, and the bags washed for a little time in cold water. On short feed, they rarely give much trouble in this respect. When thoroughly dried off, they should have the best fare, to enable them to recover condition for subsequent breeding and wintering. The fall is a critical period in which to lose flesh, either for sheep or lambs; and if any are found deficient, they should at once be provided with extra feed and attention. If cold weather overtake them, poor or in ill health, they will scarcely outlive it; or if by chance they survive, their emaciated carcass, impaired constitution, and scant fleece will ill repay the food and attention they will have cost.

CASTRATION AND DOCKING.

Some breeders advocate castration in a day or two after birth, while others will not allow the operation to be performed until the lamb is a month old. The weight of authority, however, is in favor of any time between two and six weeks after birth, when the creature has attained some strength, and the parts have not become too rigid. In such circumstances, the best English breeders recommend from ten to fifteen days old as the proper time. A lamb of a day old cannot be confirmed in all the functions of its body, and, indeed, in many instances, the testicles can then scarcely be found. At a month old, on the other hand, the lamb may be so fat, and the weather so warm, that the operation may be attended with febrile action. Dry, pleasant weather should be selected for this: a cool day, if possible; if warm, it should be done early in the morning.

Castration is a simple and safe process. Let a man hold a lamb with its back pressed firmly against his breast and stomach, and all four legs gathered in front in his hands. Cut off the bottom of the pouch, free the testicle from the inclosing membrane, and then draw it steadily out, or clip the cord with a knife if it does not snap off at a proper distance from the testicle. Some shepherds draw both testicles at once with their teeth. It is usual to drop a little salt into the pouch. Where the weather is very warm, some touch the end of the pouch with an ointment, consisting of tar, lard, and turpentine. As a general thing, however, the animal will do as well without any application.

The object of *docking* is to keep the sheep behind clean from filth and vermin; since the tail, if left on, is apt to collect filth, and, if the animal purges, becomes an intolerable nuisance. The tail, however, should not be docked too short, since it is a protection against cold in winter. This operation is by many deferred till a late period, from apprehension of too much loss of blood; but, if the weather be favorable and the lamb in good condition, it may be performed at the same time as castration with the least trouble and without injury.

The tail should be laid upon a plank, the animal being held in the same position as before. With one hand the skin is drawn toward the body, while another person, with a two-inch chisel and mallet, strikes it off at a blow, between the bone-joints, leaving it from one and a half to two inches long. The skin immediately slips back over the wound, which is soon healed. Should bleeding continue—as, however, rarely happens—so long as to sicken the lamb, a small cord should be tied firmly round the end of the tail; but this must not be allowed to remain on above twenty-four hours, as the points of the tail would slough off. Ewe lambs should be docked closer than rams. To prevent flies and maggots, and assist in healing, it is well to apply an ointment composed of lard and tar, in the proportion of four pounds of the former to one quart of the latter. The lambs should be carefully protected from cold and wet till they are perfectly well.

FEEDING.

As soon as the warm weather approaches and the grass appears, sheep become restive and impatient for the pasture. This instinct should be repressed till the ground has become thoroughly dry, and the grass has acquired substance. They ought, moreover, to be provided for the change of food by the daily use of roots for a few days before turning out. The tendency to excessive purging which is induced by the first spring-feed, may be checked by housing them at night and feeding them for the first few days with a little sound, sweet hay. They must be provided with pure water and salt; for, though they may do tolerably well without either, yet thrift and freedom from disease are cheaply secured by this slight attention.

As to *water*, it may be said that it is not indispensable in the summer pastures, since the dews and the succulence of the feed answer as a substitute; but a wide experience having demonstrated that free access to it is advantageous, particularly to those having lambs, it should be considered a matter of importance on a sheep-farm so to arrange the pastures, if possible, as to bring water into each of them.

SALT is indispensable to the health, especially in the summer. It is common to give it once a week, while they are at grass. It is still better to give them free access to it, at all times, by keeping it in a covered box, open on one side, as in the engraving annexed. A large hollow log, with holes cut along the side for the insertion of the heads of the animals, answers very well. A sheep having free access to salt at all times will never eat too much of it; and it will take its supply at such times and in such quantities as Nature demands, instead of eating of it voraciously at stated periods, as intermediate abstinence will stimulate it to do When salt is fed but once a week, it is better to have a stated day, so

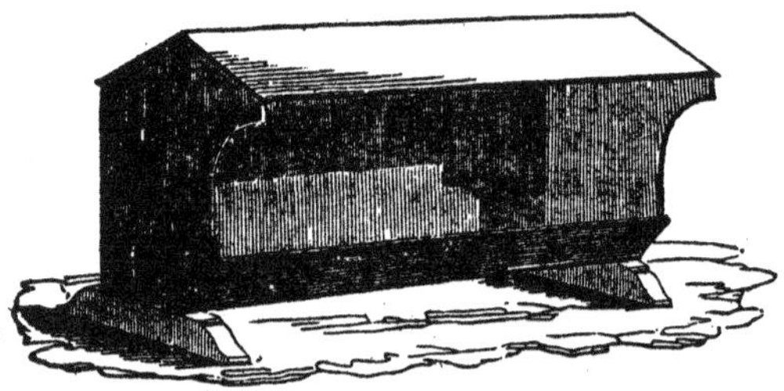

A COVERED SALTING BOX.

that it will not be forgotten; and it is well to lay the salt on flat stones—though if laid in little handfuls on the grass, very little of it will be lost.

TAR. This is supposed by many to form a very healthful condiment for sheep, and they smear the nose with it, which is licked and swallowed as the natural heat of the flesh, or that of the weather, causes it to trickle down over the nostrils and lips. Others, suffering the flock to get unusually salt-hungry, place tar upon flat stones, or in troughs, and then scatter salt upon it so that both may be consumed together. Applied to the nose, in the nature of a cataplasm, it may be advantageous in catarrhs; and in the same place, at the proper periods, its odor may, perhaps, repel the fly, the eggs of which produce the "gout in the head," as it is termed. However valuable it may be as a medicine, and even as a debergent in the case specified, there is but slight ground for confidence in it merely as a condiment.

Dry, sweet pastures, and such as abound in aromatic and bitter plants, are best suited for sheep-walks. No animal, with the exception of the goat, crops so great a variety of plants. They eat many which are rejected by the horse and the ox, which are even essential to their own wants. In this respect they are valuable assistants to the husbandman, as they feed greedily on wild mustard, burdock, thistles, marsh-mallows, milk-weed, and various other offending plants; and the Merino exceeds the more recent breeds in the range of his selections.

In pastures, however, where the dry stalks of the burdock, or the hound's-tongue, or tory-weed have remained standing over the winter, the burs are caught in the now long wool, and, if they are numerous, the wool is rendered entirely unmarketable and almost valueless. Even the dry prickles of the common and Canada thistles, where they are very numerous, get into the neck-wool of sheep, as they thrust their heads under and among them to crop the first scarce feed of the northern spring; and, independently of injuring the wool, they make it difficult to wash and otherwise handle the sheep. Indeed, it is a matter of the soundest policy to keep sheep on the cleanest pastures, those free from these and similar plants; and in a region where they are pastured the year round, they should be kept from contact with them for some months prior to shearing.

Many prepare *artificial pastures* for their flocks, which may be done with a number of plants. Winter rye, or wheat sown early in the season, may be fed off in the fall, without injury to the crop; and, in the following spring, the rye may be pastured till the stalks shoot up and begin to form a head. This affords an early and nutritious food. Corn may be sown broadcast, or thickly in drills, and either fed off in the fields or cut and carried to the sheep in their folds. White mustard is also a valuable crop for this purpose.

To give sheep sufficient variety, it is better *to divide their range* into several smaller ones, and change them as often, at least, as once a week. They seek a favorite resting-place, on a dry, elevated part of the field, which soon becomes soiled. By removing them from this for a few days, rain will cleanse or the sun dry it, so as to make it again suitable for them. More sheep may be kept, and in better condition, where this practice is adopted, than where they are confined to the same pasture.

SHADE. No one who has observed with what eagerness sheep seek shade in hot weather, and how they pant and apparently suffer when a hot sun is pouring down upon their nearly naked bodies, will doubt that, both as a matter of humanity and utility, they should be provided, during the hot summer-months, with a better shelter than that afforded by a common rail-fence. Forest trees are the most natural and the best shades, and it is as contrary to utility as it is to good taste to strip them entirely from the sheep-walks. A strip of stone-wall or close board fence on the south and west sides of the pasture, forms a tolerable substitute for trees. But in the absence of all these and of buildings of any kind, a shade can be cheaply constructed of poles and brush, in the same manner as the sheds of the same materials for winter shelter, which will be hereafter described.

FENCES. Poor *fences* will teach ewes and wethers, as well as rams, to jump; and for a jumping flock there is no remedy but immoderately high fences, or extirpation. One jumper will soon teach the trick to a whole flock; and if one by chance is brought in, it should be immediately hoppled or killed. The last is by far the surest and safest remedy.

HOPPLING is done by sewing the ends of a leather strap, broad at the extremities, so that it will not cut into the flesh, to a fore and hind leg, just above the pastern joints, leaving the legs at about the natural distance apart. *Clogging* is fastening a billet of wood to the fore leg by a leather strap. *Yoking* is fastening two rams two or three feet apart, by bows around their necks, inserted in a light piece of timber, some two or three inches in size. *Poking* is done by inserting a bow in a short bit of light timber, into which bit—worn on the under side of the neck—a rod is inserted, which projects a couple of feet in front of the sheep.

These and similar devices, to prevent rams from scaling fences, may be employed as a last resort by those improvident farmers who prefer, by such troublesome, injurious, and, at best, insecure means, to guard against that viciousness which they might so much more easily have prevented from being acquired.

DANGEROUS RAMS. From being teased and annoyed by boys, or petted and played with when young, and sometimes without any other stimulant than a naturally vicious temper, rams occasionally become very troublesome by their propensity to attack men or cattle. Some will allow no man to enter the field where they are without making an immediate onset upon him; while others will knock down the ox or horse which presumes to dispute a lock of hay with them. A ram which is known to have acquired this propensity should at once be *hooded*, and, if not valuable, at the proper season converted into a wether. But the courage thus manifested is usually the concomitant of great strength and vigor of constitution, and of a powerfully developed frame. If good in other particulars, it is a pity to lose the services of so valuable an animal. In such cases, they may be hooded, by covering their faces with leather in such a manner that they can only see a little backward and forward. They must then, however, be kept apart from the flock of rams, or they will soon be killed or injured by blows, which they cannot see to escape.

It sometimes happens that a usually quiet-tempered ram will suddenly exhibit some pugnacity when one is salting or feeding the flock. If such a person turns to run, he is immediately knocked down, and the ram learns, from that single lesson, the secret of his mastery, and the propensity to exercise it. As the ram gives his blow from the summit of the parietal and the posterior portion of the frontal bones on *the top* of his head, and not from the forehead, he is obliged to crouch his head so low when he makes his onset that he does not see forward well enough to swerve suddenly from his right line, and a few quick motions to the right and left enable one to escape him. Run in upon him, as he dashes by, with pitch-fork, club, or boot-heel, and punish him severely by blows about the head, if the club is used, giving him no time to rally until he is thoroughly cowed. This may be deemed harsh treatment, and likely to increase the viciousness of the animal. Repeated instances have, however, proved the contrary; and if the animal once is forced to acknowledge that he is overcome, he never forgets the lesson.

Prairie feeding. Sheep, when destined for the prairies, ought to commence their journey as early after the shearing as possible, since they are then disencumbered of their fleece, and do not catch and retain as much dust as when driven later; feed is also generally better, and the roads are dry and hard. Young and healthy sheep should be selected, with early lambs; or, if the latter are too young, and the distance great, they should be left, and the ewes dried off. A large wagon ought to accompany the flock, to carry such as occasionally give out; or they may be disposed of whenever they become enfeebled. With good care, a hardy flock may be driven at the rate of twelve or fourteen miles a day. Constant watchfulness is requisite, in order to keep them healthy and in good plight. One-half the expense of driving may be saved by the use of well-trained shepherd-dogs.

When arrived at their destination, they must be thoroughly washed, to free them from all dirt, and closely examined as to any diseases which they may have contracted, that these may be promptly removed. A variety of suitable food and good shelter must be provided for the autumn, winter, and spring ensuing, and every necessary attention given to them. This would be necessary if they were indigenous to the country; but it is much more so when they have just undergone a campaign to which neither they nor their race have been accustomed.

Sheep cannot be kept on the prairies without much care, artificial food, and proper attention; and losses have often occurred, by reason of a false system of economy attempted by many, from disease and mortality in the flocks, amply sufficient to have made a generous provision for the comfort and security of twice the number lost. More especially do they require proper food and attention after the first severe frosts set in, which wither and kill the natural grasses. By nibbling at the bog—the frostbitten, dead grass—they are inevitably subject to constipation, which a bountiful supply of roots, sulphur, etc., is alone sufficient to remove.

Roots, grain, good hay, straw, corn-stalks, and pea or bean-vines are essential to the preservation of their health and thrift during the winter, everywhere north of thirty-nine degrees. In summer, the natural herbage is sufficient to sustain them in fine condition, till they shall have acquired a denser population of animals, when it will be found necessary to stock their meadows with the best varieties of artificial grasses.

The prairies seem adapted to the usual varieties of sheep introduced into the United States; and of such are the flocks made up, according to the taste or judgment of the owners. Shepherd dogs are invaluable to the owners of flocks, in these unfenced, illimitable ranges, both as a defence against the small prairie wolves, which prowl around the sheep, but have been rapidly thinned off by the settlers, and also as assistants to the shepherds in driving and herding their flocks on the open ground.

FALL FEEDING. In the North, the grass often gets very short by the tenth or fifteenth of November, and it has lost most of its nutritiousness from repeated freezing and thawing. At this time, although no snow may have fallen, it is best to give the sheep a light, daily foddering of bright hay, and a few oats in the bundle. Given thus for the ten or twelve days which precede the covering of the ground by snow, fodder pays for itself as well as at any other time during the year. It is well to feed oats in the bundle, or threshed oats, about a gill to the head, in the feeding-troughs, carried to the field for that purpose.

WINTER FEEDING. The time for taking sheep from the pastures must depend on the state of the weather and food. Severe frosts destroy much of the nutriment in the grasses, and they soon after cease to afford adequate nourishment. Long exposure to cold storms, with such food to sustain them, will rapidly reduce the condition of these animals. The only safe rule is to transfer them to their winter-quarters the first day they cease to thrive abroad.

There is no better food for sheep than well-ripened, sound Timothy hay; though the clovers and nearly all the cultivated grasses may be advantageously fed. Hundreds and thousands of northern flocks receive, during the entire winter, nothing but ordinary hay, consisting mainly of Timothy, some red and white clover, and frequently a sprinkling of gum, or spear grass. Bean and pea straw are valuable, especially the former, which, if properly cured, they prefer to the best hay; and it is well adapted to the production of wool. Where hay is the principal feed, it may be well, where it is convenient, to give corn-stalks every fifth or sixth feed, or even once a day; or the daily feed, not of hay, might alternate between stalks, pea-straw, straws of the cereal grains, etc. It is mainly a question of convenience with the farmer, provided a proper supply of palatable nutriment within a proper compass is given. It would not, however, be entirely safe to confine any kind of sheep to the straw of the cereal grains, unless it were some of those little hardy varieties of animals which would be of no use in this country.

The expediency of feeding *grain* to store-sheep in winter depends much on circumstances. If in a climate where they can obtain a proper supply of grass or other green esculents, it would, of course, be unnecessary; nor is it a matter of necessity where the ground is frozen or covered with snow for weeks or months, provided the sheep be plentifully supplied with good dry fodder. Near markets where the coarser grains find a quick sale at fair prices, it is not usual, in the North, to feed grain. Remote from markets it is generally fed by the holders of large flocks. Oats are commonly preferred, and they are fed at the rate of a gill a head per day. Some feed half the same amount of yellow corn. Fewer sheep, particularly lambs and yearlings, get thin and perish where they receive a daily feed of grain; they consume less hay, and their fleeces are increased in weight. On the whole, therefore, it is considered good economy. Where no grain is fed, three daily feeds of hay are given. The smaller sizes of the Saxon may be well sustained on two pounds of hay; but larger sheep will consume from three and a half to four or even five pounds per day. Sheep, in common with all other animals, when exposed to cold, will consume much more than if well protected, or during a warmer season.

It is a common and very good practice to feed greenish cut oats in the bundle, at noon, and give but two feeds of hay, one at morning and one at night. Some feed greenish cut peas in the same way. In warm, thawing weather, when sheep get to the ground and refuse dry hay, a little grain assists materially in keeping up their strength and condition. When the feed is shortest in winter, in the South, there are many localities where sheep can get enough grass to take off their appetite for dry hay, but not quite enough to keep them in prime order. A moderate daily feed of oats or pease, placed in the depository racks, would keep them strong and in good plight for the lambing season, and increase their weight of wool.

Few Northern farmers feed *Indian corn* to store-sheep, as it is considered too hot and stimulating, and sheep are thought to become more liable to become "cloyed" on it than on oats, pease, etc. Yellow corn is not generally judged a very safe feed for lambs and yearlings. Store-sheep should be kept in good, fair, plump condition. Lambs and yearlings may be as fat as they will become on proper feeding. It is stated that sheep will eat *cotton-seed*, and thrive on it.

It must be remembered that sheep are not to be allowed to get thin during the winter, with the idea that their condition can at any time be readily raised by better feed, as with the horse or ox. It is always difficult, and, unless properly managed, expensive and hazardous, to attempt to raise the condition of a poor flock in the winter, especially if they have reached that point where they manifest weakness. If the feeding of a liberal allowance of grain be suddenly commenced, fatal diarrhœa will often supervene. All extra feeding, therefore, must be begun very gradually; and it does not appear, in any case, to produce proportionable results.

Roots, such as ruta-bagas, Irish potatoes, and the like, make a good substitute for grain, or as extra feed for grown sheep. The ruta-baga is preferable to the potatoe in its equivalents of nutriment. No root, however, is as good for lambs and yearlings as an equivalent of grain. Sheep may be taught to eat nearly all the cultivated roots. This is done by withholding salt from them, and then feeding the chopped roots a few times, rubbed with just sufficient salt to induce them to eat the root to obtain it; but not enough to satisfy their appetite for salt before they have acquired a taste for the roots.

It is customary with some farmers to cut down, from time to time in the winter, and draw into the sheep-yards, young trees of the *hemlock*, whose foliage is greedily eaten by the sheep, after being confined for some time to dry feed. This browse is commonly used, like tar, for some supposed medicinal virtues. It is pronounced "healthy" for sheep. Much the same remarks might be made about this as have been already made concerning tar. No tonics and stimulants are needed for a healthy animal. If the foliage of the hemlock were constantly accessible to them, there would be no possible objection to their eating it, since their instincts, in that case, would teach them whether, and in what quantities, to devour it; but when entirely confined to dry feed for a protracted period, sheep will consume injurious and even poisonous succulents, and of the most wholesome ones, hurtful quantities. As a mere *laxative*, an occasional feed of hemlock may be beneficial; though, in this point of view, a day's run at grass, in a thaw, or a feed of roots, would produce the same result. In a climate where grass is procurable most of the time, browse for medicinal purposes is entirely unnecessary.

Sheep undoubtedly require *salt* in winter. Some salt their hay when it is stored in the barn or stack. This is objectionable, since the appetite of the sheep is much the safest guide in the premises. It may be left accessible to them in the salt-box, as in summer; or an occasional feed of grined hay or straw may be given them in warm, thawing weather, when their appetite is poor. This last is an excellent plan, and serves a double purpose. With a wisp of straw, sprinkle a thin layer of straw with brine, then another layer of straw, and another sprinkling, and so on. Let this lie until the next day, for the brine to be absorbed by the straw, and then feed it to all the grazing animals on the farm which need salting.

Water is indispensable, unless sheep have access to succulent food, or clean snow. Constant access to a brook or spring is best; but, in default of this, they should be watered at least *once a day* in some other way.

Feeding with other stock. Sheep should not run, or be fed, *in yards*, with any other stock. Cattle hook them, often mortally; and colts tease and frequently injure them. It is often said that "colts will pick up what sheep leave." But well-managed sheep rarely leave any thing; and, if they chance so to do, it is better to rake it up and throw it into the colts' yard, than to feed them together. If sheep are not required to eat their food pretty clean, they will soon learn to waste large quantities. If, however, they are overfed with either hay or grain, it is not proper to compel them, by starvation, to come back and eat it. This they will not do, unless sorely pinched. Clean out the troughs, or rake up the hay, and the next time feed less.

Division of flocks. If flocks are shut up in small inclosures during winter, according to the northern custom, it is necessary to divide them into flocks of about one hundred each, consisting of sheep of about the same size and strength; otherwise, the stronger rob the weaker, and the latter rapidly decline. This is not so important where the sheep roam at large; but, even in that case, some division and classification are best. It is best, indeed, even in summer. The poorer and feebler can by this means receive better pasture, or a little more grain and better shelter in winter.

By those who grow wool to any extent, breeding ewes, lambs, and wethers, are invariably kept in separate flocks in winter; and it is best to keep yearling sheep by themselves with a few of the smallest two-year-olds, and any old crones which are kept for their excellence as breeders, but which cannot maintain themselves in the flock of breeding ewes.

Old and feeble or wounded sheep, late-born lambs, etc.,

should be placed by themselves, even if the number be small, as they require better feed, warmer shelter, and more attention. Unless the sheep are of a peculiarly valuable variety, however, it is better to sell them off in the fall at any price, or to give them to some poor neighbor who has time to nurse them, and who may thus commence a flock.

REGULARITY IN FEEDING If any one principle in sheep husbandry deserves careful attention more than others, it is, that *the utmost regularity must be preserved in feeding.*

First, there should be regularity as to *the times* of feeding. However abundantly provided for, when a flock are foddered sometimes at one hour and sometimes at another—sometimes three times a day, and sometimes twice—some days grain, and some days none—they cannot be made to thrive. They will do far better on inferior keep, if fed with strict regularity. In a climate where they require hay three times a day, the best times for feeding are about sunrise in the morning, at noon, and an hour before dark at night. Unlike cattle and horses, sheep do not feed well in the dark; and, therefore, they should have time to consume their food before night sets in. Noon is the common time for feeding grain or roots, and is the best time, if but two fodderings of hay are given. If the sheep receive hay three times, it is not a matter of much consequence with which feeding grain is given, only that the practice be uniform.

Secondly, it is highly essential that there should be regularity in *the amount* fed. The consumption of hay will, it is true, depend much upon the weather; the keener the cold, the more the sheep will eat. In the South, much depends upon the amount of grass obtained. In many places, a light, daily foddering supplies; in others, a light foddering placed in the depository racks once in two days, answers the purpose. In the steady cold weather of the North, the shepherd readily learns to determine about how much hay will be consumed before the next foddering time. And this amount should, as near as may be, be regularly fed. In feeding grain or roots, there is no difficulty in preserving entire regularity; and it is vastly more important than in feeding hay. Of the latter, a sheep will not over-eat and surfeit itself; of the former, it will. Even if it be not fed grain to the point of surfeiting, it will expect a like amount, however over-plenteous, at the next feeding; failing to receive which, it will pine for it, and manifest uneasiness. The effect of such irregularity on the stomach and system of any animal is bad; and the sheep suffers more from it than any other animal. It is much better that the flock receive no grain at all, than that they receive it without regard to regularity in the amount. The shepherd should *measure* out the grain to the sheep in all instances, instead of *guessing* it out, and measure it to each separate flock.

EFFECT OF FOOD. Well-fed sheep, as has been previously remarked, produce more wool than poorly fed ones. No doctrine is more clearly recognized in agricultural chemistry than that animal tissues derive their chemical components from the same components existing in their food. Various analyses show that the chemical composition of wool, hair, hoofs, nails, horns, feathers, lean meat, blood, cellular tissue, nerves, etc., are nearly identical.

The organic part of wool, according to standard authorities, consists of carbon, 50.65; hydrogen, 7.03; nitrogen, 17.71; oxygen and sulphur, 24.61. The inorganic constituents are small. When burned, it leaves but a trifling per cent. of ash.

The large quantity of nitrogen contained in wool shows that its production is increased by highly azotized food; and from various experiments made, a striking correspondence has been found to exist between the amount of wool and the amount of nitrogen in food. *Pease* rank first in increasing the wool, and very high in the average comparative increase which they produce in all the tissues.

The increase of fat and muscle, as of wool, depends upon the nature of the food. It is not very common, in the North, for wool-growers to fatten their wethers for market by extra winter feeding. Some give them a little more generous keep the winter before they are to be turned off, and then salt them when they have obtained their maximum fatness the succeding fall.

Stall-feeding is lost on an ill-shaped, unthrifty animal. The perfection of form and health, and the uniform good condition which characterizes the thrifty one, indicate, too plainly to be misunderstood, those which will best repay the care of their owner. The selection of any indifferent animal for stall-fattening will inevitably be attended with loss. Such ought to be got rid of, when first brought from the pasture, for the wool they will bring.

When winter fattening is attempted, sheep require warm, dry shelters, and should receive, in addition to all the hay they will eat, meal twice a day in troughs—or meal once and chopped roots once. The equivalent of from half a pint to a pint of yellow corn meal per head each day is about as much as ordinary stocks of Merino wethers will profitably consume; though in selected flocks, consisting of large animals, this amount is frequently exceeded.

YARDS.

Experience has amply demonstrated that—in the climate of the Northern and Eastern States, where no grass grows from four to four and a half months in the winter, and where, therefore, all that can be obtained from the ground is the repeatedly frozen, unnutritious herbage left in the fall—it is better to keep sheep confined in yards, excepting where the ground is covered with snow. If suffered to roam over the fields at other times, they get enough grass to take away their appetite for dry hay, but not enough to sustain them; they fall away, and toward spring they become weak, and a large proportion of them frequently perish. Flocks of some size are here, of course, alluded to, and on properly stocked farms. A few sheep would do better with a boundless range.

Some let out their sheep occasionally for a single day, during a thaw; others keep them entirely from the ground until let out to grass in the spring. The former course is preferable where the sheep ordinarily get nothing but dry fodder. It affords a healthy laxative, and a single day's grazing will not take off their appetite from more than one succeeding dry feed. It is necessary, in the North, to keep sheep in the yards until the feed has got a good start in the spring, or they will get off from their feed—particularly the breeding ewes—and get weak at the most critical time for them in the year.

Yards should be firm-bottomed, dry, and, in the northern climate, kept well littered with straw. The yarding system is not practised to any great extent in the South; nor should it be, where sheep can get their living from the fields.

FEEDING-RACKS.

When the ground is frozen, and especially when covered with snow, the sheep eats hay well on the ground; but when the land is soft, muddy, or foul with manure, they will scarcely touch hay placed on it—or, if they do, will tread much of it

A CONVENIENT BOX-RACK.

into the mud, in their restlessness while feeding. It should then be fed in racks, which are more economical, even in the first-named case; since, when the hay is fed on the ground, the leaves and seeds, the most valuable part of the fodder, are almost wholly lost.

To make an economical *box-rack*—the one in most general use in the North—take six light pieces of scantling, say three inches square, one for each corner, and one for the centre of each side. Boards of pine or hemlock, twelve or fifteen feet long, and twelve or fourteen inches wide, may then be nailed on to the bottom of the posts for the sides, which are separated by similar boards at the ends, two and a half feet long. Boards twelve inches wide, raised above the lower ones by a space of from nine to twelve inches, are nailed on the sides and ends, which completes the rack. The edges of the opening should be made perfectly smooth, to prevent chafing or tearing out the wool. The largest dimensions given are suitable for the large breeds, and the smallest for the Saxon; and still smaller are proper for the lambs. These should be set on dry ground, or under the sheds; and they can be easily removed wherever necessary. Unless over-fed, sheep waste very little hay in them.

Some prefer the racks made with slats, or smooth, upright sticks, in the form of the common horse-rack. This kind should always be accompanied by a broad trough affixed to the bottom, to catch the fine hay which falls in feeding These racks may be attached to the side of a building, or used double. A small lamb requires fifteen inches of space, and a large sheep two feet, for quiet, comfortable feeding; and this amount of room, at least, should be provided around the racks for every sheep.

With what is termed a *hole-rack*, sheep do not crowd and take advantage of each other so much as with log-racks; but

A HOLE-RACK.

they are too heavy and unnecessarily expensive for a common out-door rack. This rack is box-shaped, with the front formed of a board nailed on horizontally, or, more commonly, by nailing the boards perpendicularly, the bottoms on the sill of a barn, and the tops to horizontal pieces of timber. The holes should be at least eight inches wide, nine inches high, and eighteen inches from centre to centre.

In the South, racks are not so necessary for that constant use to which they are put in colder sections, as they are for depositories of dry food, for the occasional visitation of the sheep. In soft, warm weather, when the ground is unfrozen, and any kind of green herbage is to be obtained, sheep will scarcely touch dry fodder; though the little they will then eat will be highly serviceable to them. But in a sudden freeze, or on the occurrence of cold storms, they will resort to the racks, and fill themselves with dry food. They anticipate the coming storm by instinct, and eat an extra quantity of food to sustain the animal heat during the succeeding depression of temperature. They should always have racks of dry fodder for resort in such emergencies.

These racks should have covers or roofs to protect their contents from rain, as otherwise the feed would often be spoiled before but a small portion of it would be consumed. Hay or straw, saturated with water, or soaked and dried, is only eaten by the sheep as a matter of absolute necessity. The common box-rack would answer the purpose very well by placing on the top a triangular cover or roof, formed of a couple of boards, one hung at the upper edge with iron or leather hinges, so that it could be lifted up like a lid; making the ends tight; drawing in the lower edges of the sides, so that it should not be more than a foot wide on the bottom; inserting a flow; and then mounting it on, and making it fast to, two cross-sills, four or five inches square, to keep the floor off from the ground, and long enough to prevent it from being easily overturned. The lower side-board should be narrow, on account of the increased height given its upper edge by the sills.

A rack of the same construction, with the sides like those described for the hole-rack, would be still better, though somewhat more expensive; or the sides might consist of rundles, the top being nailed down in either case, and the fodder inserted by little doors in the ends.

What is termed the *hopper-rack*, serving both for a rack and a feeding-trough, is a favorite with many sheep-owners. The accompanying cut represents a section of such a rack. A piece of durable wood, about four and a half feet long, six or eight inches deep, and four inches thick, having two notches, *a a*, cut into it, and two troughs, made of inch boards, *b b b b*, placed in these notches, and nailed fast, constitute the formation. If the rack is to be fourteen feet long, three sills are required. The ends of the rack are made by nailing against the side of the sill-boards that reach up as high as it is desired to have the rack; and nails driven through these end-boards into the ends of the side-boards, *f f*, secure them. The sides may be further strengthened by pieces of board on the outside of them, fitted into the trough. A roof may be put over all, if desired, by means of which the fodder is kept entirely from the weather, and no seeds or chaff can get into the wool.

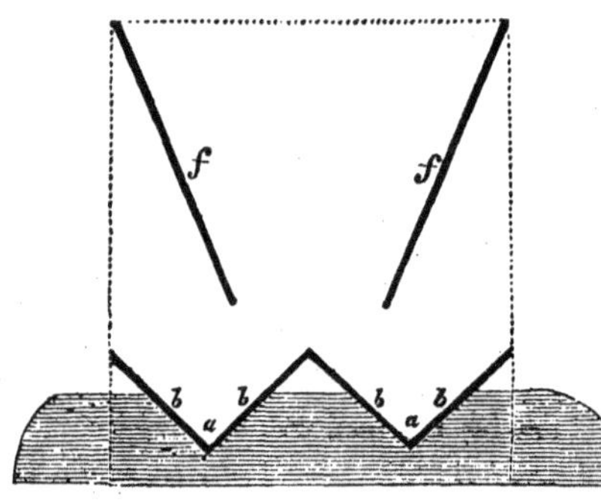

THE HOPPER-RACK.

TROUGHS.

Threshed grain, chopped roots, etc., when fed to sheep, should be placed in troughs. With either of the racks which have been described, except the last, a separate trough would be required. The most economical are made of two boards of any convenient length, ten to twelve inches wide. Nail the lower side of one upon the edge of the other, fastening both into a two or three-inch plank, fifteen inches long, and a foot

AN ECONOMICAL SHEEP-TROUGH.

wide; notched in its upper edge in the form required. In snowy sections they are turned over after feeding, and when falls of snow are anticipated one end is laid on the yard-fence.

Various contrivances have been brought to notice for keeping grain where sheep can feed on it at will, a description of which is omitted, since it is not thought best, by the most successful stock-raisers, in feeding or fattening any quadrupeds, to allow them grain at will, stated feeds being preferred by them; and the same is true of fodder. If this system is departed from in using depository racks, as recommended, it is because it is rendered necessary by the circumstances of the case. A Merino store-sheep, allowed as much grain as it chose to consume, would be likely to inflict injury on itself; and grain so fed would, generally speaking, be productive of more damage than benefit.

BARNS AND SHEDS.

Shelters, in northern climates, are indispensable to profitable sheep-raising; and in every latitude north of the Gulf of Mexico, they would probably be found advantageous. An animal eats much less when thus protected; he is more thrifty, less liable to disease, and his manure is richer and more abundant. The feeding may be done in the open yard in clear weather, and under cover in severe storms: for, even in the vigorous climate of the North, none but the breeders of Saxons make a regular practice of feeding under cover.

Humanity and economy alike dictate that, in the North, sheep should be provided with shelters under which to lie

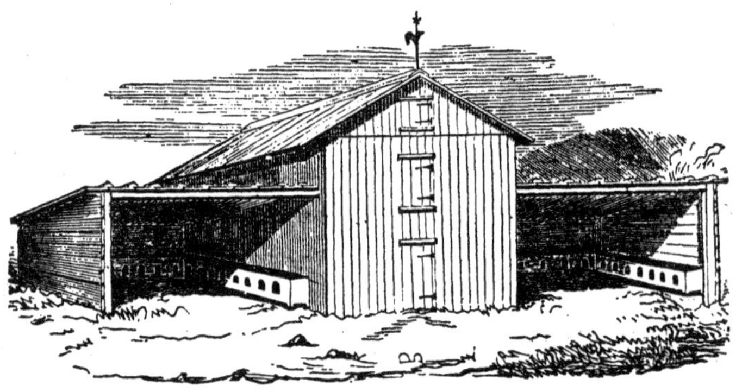

SHEEP-BARN WITH SHEDS.

nights, and to which they can resort at will. It is not an uncommon circumstance in New York and New England for snow to fall to the depth of from twenty to thirty inches within twenty-four or forty-eight hours, and then to be succeeded by a strong and intensely cold west or northwest wind of several days continuance, which lifts the snow, blocking up the roads, and piling huge drifts to the leeward of fences, barns, etc.

A flock without shelter will huddle closely together, turning their backs to the storm, constantly stepping, and thus treading down the snow as it rises about them. Strong, close-coated sheep do not seem to suffer as much from the cold, for a period, as would be expected. It is, however, almost impossible to feed them enough, or half enough, under such circumstances, without an immense waste of hay—entirely impossible, indeed, without racks. The hay is whirled away in an instant by the wind; and, even if racks are used, the sheep, leaving their huddle, where they were kept warm and even moist by the melting snow in their wool, soon get chilled, and are disposed to return to their huddle. Imperfectly filled with food, the supply of animal heat is lowered, and, at the end of the second or third day, the feeble ones sink down hopelessly, the yearlings, and those somewhat old, receive a shock from which nothing but the most careful nursing will enable them to rally, and even the strongest suffer an injurious loss in condition.

Few persons, therefore, who own as many as forty or fifty sheep, attempt to get along without some kind of shelters, which are variously constructed, to suit their tastes or circumstances. A sheep-barn, built upon a side-hill, will afford two floors: one underneath, surrounded by three sides of wall, should open to the south, with sliding or swinging doors to guard against storms; and another may be provided above, if the floors are perfectly tight, with proper gutters to carry off the urine; and sufficient storage for the fodder can be furnished by scaffolds overhead. They may also be constructed with twelve or fifteen-feet posts on level ground, allowing the sheep to occupy the lower part, with the fodder stored above.

In all cases, however, *thorough ventilation should be provided;* for of the two evils, of exposure to cold or of too great privation of air, the former is to be preferred. Sheep cannot long endure close confinement without injury. In all ordinary weather, a shed, closely boarded on three sides, with a light roof, is sufficient protection; especially if the open side is shielded from bleak winds, or leads into a well-inclosed yard. If the floors above are used for storage, they should be made tight, that no hay, chaff, or dust can fall upon the fleece. The sheds attached to the barn are not usually framed or silled, but are supported by some posts of durable timber set in the ground. The roofs are formed of boards battened with slats. The barn has generally no partitions within, and is entirely filled with hay.

There are many situations in which open sheds are very liable to have snow drifted under them by certain winds, and they are subject in all severe gales to have the snow carried over them to fall down in large drifts in front, which gradually encroach on the sheltered space, and are very inconvenient, particularly when they thaw. For these reasons, many prefer sheep-houses covered on all sides, with the exception of a wide doorway for ingress and egress, and one or two windows for the necessary ventilation. They are convenient for yarding sheep, and the various processes for which this is required; as for shearing, marking, sorting, etc., and especially so for lambing-places, or the confinement of newly-shorn sheep in cold storms. They should have so much space that, in addition to the outside racks, others can be placed temporarily through the middle when required.

The facts must not be overlooked—as bearing upon the question of shelter, even in the warmer regions of the country—that cold rains, or rains of any temperature, when immediately succeeded by cold or freezing weather, or cold, piercing winds, are more hurtful to sheep than even snow-storms; and that, consequently, sheep must be adequately guarded against them.

SHEDS. The simplest and cheapest kind of shed is formed by poles or rails, the upper end resting on a strong horizontal

A SHED OF RAILS.

pole supported by crotched posts set in the ground. It may be rendered rain-proof by pea-vines, straw, or pine boughs. In a region where timber is very cheap, planks or boards, of a sufficient thickness not to spring downward, and thus open the roof, battened with slats, may take the place of the poles and boughs; and they would make a tighter and more durable roof. If the lower ends of the boards or poles are raised a couple of feet from the ground, by placing a log under them, the shed will shelter more sheep.

These movable sheds may be connected with hay-barns—"hay-barracks"—or they may surround an inclosed space with a stack in the middle. In the latter case, the yard should be square, on account of the divergence in the lower ends of the boards or poles, which a round form would render necessary.

Sheds of this description are frequently made between two stacks. The end of the horizontal supporting-pole is placed on the stack-pens when the stacks are built, and the middle is propped by crotched posts. The supporting-pole may rest, in the same way, on the upper girts of two hay-barracks; or two such sheds, at angles with each other, might form wings to this structure.

On all large sheep-farms, convenience requires that there be one barn of considerable size, to contain the shearing-floor, and the necessary conveniences about it for yarding the sheep, etc. This should also, for the sake of economy, be a hay-barn, where hay is used. It may be constructed in the corner of four fields, so that four hundred sheep can be fed from it, without racking flocks of improper size. At this barn it would be expedient to make the best shelters, and to bring together all the breeding-ewes on the farm, if their number does not exceed four hundred. The shepherd would thus be saved much travel at all times, and particularly at the lambing-time, and each flock would be under his almost constant supervision.

The size of this barn is a question to be determined entirely by the climate. For large flocks of sheep, the storage of some hay or other fodder for winter is an indispensable precautionary measure, at least in any part of the United States; and, other things being equal, the farther north, or the more elevated the land, the greater would be the amount necessary to be stored.

HAY-HOLDER. Where hay or other fodder is thrown out of the upper door of a barn into the sheep-yard—as it always must necessarily be in any mere hay-barn—or where it is thrown from a barrack or stack, the sheep immediately rush on it, trampling it and soiling it, and the succeeding forkfuls fall on their backs, filling their wool with dust seed, and chaff.

This is obviated by hay-holders—yards ten feet square—either portable, by being made of posts and boards, or simply a pen of rails, placed under the doors of the barns, and by the sides of each stack or barrack. The hay is pitched into this holder in fair weather, enough for a day's foddering at a time, and is taken from it by the fork and placed in the racks.

The poles or rails for stack-pens or hay-holders should be so small as to entirely prevent the sheep from inserting their heads in them after hay. A sheep will often insert his head where the opening is wide enough for that purpose, shove it along, or get crowded, to where the opening is not wide enough to withdraw the head, and it will hang there until observed and extricated by the shepherd. If, as often happens, it is thus caught when its foreparts are elevated by climbing up the side of the pen, it will continue to lose its footing in its struggles, and will soon choke to death.

TAGGING.

Tagging, or clatting, is the removal from the sheep of such wool as is liable to get fouled when the animal is turned on to the fresh pastures. If sheep are kept on dry feed through the winter, they will usually purge, more or less, when let out to green feed in the spring. The wool around and below the anus becomes saturated with dung, which forms into hard pellets, if the purging ceases. Whether this take place or not, the adhering dung cannot be removed from the wool in the ordinary process of washing; and it forms a great impediment in shearing, dulling and straining the shears to cut through it, when in a dry state, and it is often impracticable so to do. Besides, it is difficult to force the shears between it and the skin, without frequently and severely wounding the latter. Occasionally, too, flies deposit their eggs under this mass of filth prior to shearing; and the ensuing swarm of maggots, unless speedily discovered and removed, will lead the sheep to a miserable death.

Before the animals are let out to grass, each one should have the wool sheared from the roots of the tail down the inside of the thighs; it should likewise be sheared from off the entire bag of the ewe, that the newly-dropped lamb may more readily find the teat, and from the scrotum, and so much space round the point of the sheath of the ram as is usually kept wet. If the latter place is neglected, soreness and ulceration sometimes ensue from the constant maceration of the urine.

An assistant should catch the sheep and hold them while they are tagged. The latter process requires a good shearer, as the wool must be cut off closely and smoothly, or the object is but half accomplished, and the sheep will have un unsightly and ridiculous appearance when the remainder of their fleeces is taken off; while, on the other hand, it is not only improper to cut the skin of a sheep at any time, but it is peculiarly so to cut that or the bag of a ewe when near lambing. The wool saved by tagging will far more than pay the expenses of the operation. It answers well for stockings and other domestic purposes, or it will sell for nearly half the price of fleece-wool.

Care should be exercised at all times in handling sheep, especially ewes heavy with lamb. It is highly injurious and unsafe to chase them about and handle them roughly; for, even if abortion, the worst consequence of such treatment, is avoided, they become timid and shy of being touched, rendering it difficult to catch them or render them assistance at the lambing period, and even a matter of difficulty to enter the cotes, in which it is sometimes necessary to confine them at that time, without having them driving about pell-mell, running over their lambs, etc. If a sheep is suddenly caught by the wool on her running, or is lifted by the wool, the skin is to a certain extent loosened from the body at the points where it is thus seized; and, if killed a day or two afterward, blood will be found settled about those parts.

When sheep are to be handled, they should be inclosed in a yard just large enough to hold them without their being crowded, so that they shall have no chance to run and dash about. The catcher should stop them by seizing them by the hind-leg just above the hock, or by clapping one hand before the neck and the other behind the buttocks. Then, not waiting for the sheep to make a violent struggle, he should throw its right arm over and about immediately back of the shoulders, place his hand on the brisket, and lift the animal on his hip. If the sheep is very heavy, he can throw both arms around it, clasp his fingers under the brisket, and lift it up against the front part of his body. He should then set it carefully on its rump upon the tagging-table, which should be eighteen or twenty inches high, support its back with his legs, and hold it gently and conveniently until the tagger has performed his duty. Two men should not be allowed to lift the same sheep together, as it will be pretty sure to receive some strain between them. A good shearer and assistant will tag two hundred sheep per day.

When sheep receive green feed all the year round—as they do in many parts of the South—and no purging ensues from eating the newly-starting grasses in the spring, tagging is unnecessary.

WASHING.

Many judicious farmers object to washing sheep, on account of its tendency to produce colds and catarrhal affections, to which this animal is particularly subject; but it cannot well be dispensed with, as the wool is always rendered more salable; and if the operation is carefully done, it need not be attended with injury.

Mr. Randall, the extensive sheep-breeder of Texas, states that he does not wash his sheep at all, for what he deems good reasons. About the middle of April, or at the time when one-half of the ewes have young lambs at their sides, and the balance about to drop, would be the only time in that region when he could wash them. At this period he would not race or worry his ewes at all, on any account; as they should be troubled as little as possible, and no advantage to the fleece from washing could compensate for the injury to the animal. In his high mountain-region, lambing-time could not prudently come before the latter part of March or April—the very period when washing and shearing must be commenced—since in February, and even up to the fifteenth or twentieth of March, there is much bad weather, and a single cold, rainy or sleety norther would carry off one-half of the lambs dropped during its continuance.

In most of that portion of the United States lying north of forty degrees, the washing is performed from the middle of May till the first of June, according to the season and climate. When the streams are hard, which is frequently the case in limestone regions, it is better to attend to it immediately after an abundant rain, which proportionately lessens the lime derived from the springs. The climate of the Southern States would admit of an earlier time. The rule should be to wait until the water has acquired sufficient warmth for bathing, and until cold rains and storms and cold nights are no longer to be expected.

The practice of a large majority of farmers is to drive their sheep to the watering-ground early in the morning, on a warm day, leaving the lambs behind. The sheep are confined on the bank of the stream by a temporary enclosure, from which they are taken, and, if not too heavy, carried into water sufficiently deep to prevent their touching bottom. They are then washed, by gently squeezing the fleece with the hands, after which they are led ashore, and as much of the water pressed out as possible before letting them go, as the great weight retained in the wool frequently staggers and throws them down.

By the best flock-masters, sheep are usually washed in vats. A small stream is dammed up, and the water taken from it in an aqueduct, formed by nailing boards together, and carried till a sufficient fall is obtained to have it pour down a couple of feet or more into the vat. The body of water, to do the work fast and well, should be some twenty-four inches wide, and five or six deep; and the swifter the current the better. The vat should be some three and a half feet deep, and large enough for four sheep to swim in it. A yard is built near the vat, from the gate of which a platform extends to and incloses the vat on three sides. This keeps the washer from standing in the water, and makes it much easier to lift the sheep in and out. The yard is built opposite the corners of two fields—

to take advantage of the angle of one of them to drive the sheep more readily into the yard, which should be large enough to contain the entire flock, if it does not exceed two hundred; and the bottom of it, as well as the smaller yard,

WASHING APPARATUS.

unless well sodded over, should be covered with coarse gravel, to avoid becoming muddy. If the same establishment is used by a number of flock-masters, gravelling will always be necessary.

As soon as the flocks are confined in the middle yard, the lambs are all immediately caught out from among them, and set over the fence into the yard to the left, to prevent their being trampled down, as often happens, by the old sheep, or straying off, if let loose. As many sheep are then driven out of the middle yard into the smaller yard to the right, as it will conveniently hold. A boy stands by the gate next to the vat, to open and shut it, or the gate is drawn together with a chain and weight, and two men, catching the sheep as directed under the head of "tagging," commence placing them in the water for the preparatory process of "wetting." As soon as the water strikes through the wool, which occupies but an instant, the sheep is lifted out and let loose. Where there are conveniences for so doing, this process may be more readily performed by driving them through a stream deep enough to compel the sheep to swim; but *swimming* the compact-fleeced, fine-woolled sheep for any length of time—as is practised with the long-wools in England—will not properly cleanse the wool for steaming. The vat should, of course, be in an inclosed field, to prevent their escape. The whole flock should thus be passed over, and again driven round through the field into the middle yard, where they should stand for about an hour before washing commences.

There is a large per centage of potash in the wool oil, which acts upon the dirt independent of the favorable effect which would result from thus soaking it with water alone for some time. If washed soon after a good shower, previous wetting might be dispensed with; and it is not, perhaps, absolutely necessary in any case. If the water is warm enough to allow the sheep to remain in it for the requisite period, they may be got clean by washing without any previous wetting; though the snowy whiteness of fleece, which has such an influence on the purchaser, is not so often nor so perfectly attained in the latter way. But little time is saved by dispensing with "wetting," as it takes proportionably longer to wash, and it is not so well for the sheep to be kept so long in the water at once.

When the washing commences, two and sometimes four sheep are plunged in the vat. When four are put in, two soak while two are washed. This should not, however, be done, unless the water is very warm, and the washers are uncommonly quick and expert; and it is, upon the whole, rather an objectionable practice, since few animals suffer so much from the effects of a chill as the sheep; and, if they have been previously wetted, it is wholly unnecessary. When the sheep are in the water, the two washers commence kneading the wool with their hands about the dirtier parts—the breech, belly, etc.—and they continue to turn the sheep so that the descending current of water can strike into all parts of the fleece.

As soon as the sheep are clean, which may be known by the water running entirely clear, each washer seizes his own animal by the foreparts, plunges it deep in the vats, and, taking advantage of the rebound, lifts it out, setting it gently down on its breech upon the platform. He then—if the sheep is old and weak, and it is well in all cases—presses out some of the water from the wool, and after submitting the sheep to a process presently to be mentioned, lets it go.

There should be no mud about the vat, the earth not covered with sod, being gravelled. Sheep should be kept on clean pastures, from washing to shearing—not where they can come in contact with the ground, burnt logs, and the like—and they should not be driven over dusty roads. The washers should be strong and capable men, and, protected as they are from any thing but the water running over the sides of the vat, they can labor several hours without inconvenience. Two hundred sheep will employ two experienced men not over half a day, and this rate is at times much exceeded.

It is a great object, not only as a matter of propriety and honesty, but even as an item of profit, to get the wool clean, and of a snowy whiteness, in which condition it will always sell for more than enough extra to offset the increased labor and the diminution in weight. The average loss in American Saxon wool in scouring, after being washed on the back, is estimated at thirty-six per cent.; and in American Merino forty-two and a half per cent.

CUTTING THE HOOFS.

As the hoofs of fine-woolled sheep grow rapidly, turning up in front and under at the sides, they must be clipped as often as once a year, or they become unsightly, give an awkward, hobbling gait to the animal, and the part of the horn which turns under at the sides holds dirt or dung in constant contact with the soles, and even prevents it from being readily shaken or washed out of the cleft of the foot in the natural movement of the sheep about the pastures, as would take place were the hoof in its proper place. This greatly aggravates the hoof-ail, and renders the curing of it more difficult; and it is thought by many to be the exciting cause of the disease.

It is customary to clip the hoofs at tagging, or at or soon after the time of shearing. Some employ a chisel and mallet to shorten the hoofs; but the animal must afterward be turned upon its back, to pare off the crust which projects and turns under. If the weather be dry, or the sheep have stood for some time on dry straw, as at shearing, the hoofs are as tough as horn, and are cut with great difficulty; and this is increased by the grit and dirt adhering to the sole, which immediately takes the edge off from the knife. These periods are ill-chosen, and the method slow and bungling. It is particularly improper to submit heavily-pregnant ewes to all this unnecessary handling at the time of tagging.

When the sheep is washed and lifted out of the vat, and placed on its rump upon the platform, the gate-keeper should advance with a pair of toe-nippers, and the washer present each foot separately, pressing the toes together so that they can be severed at a single clip. The nippers—which can be made by any blacksmith who can temper an axe or a chisel—must be made strong, with handles a little more than a foot long, the rivet being of half-inch iron, and confined with a nut, so that they may be taken apart for sharpening. The cutting-edge should descend upon a strip of copper inserted in the iron, to prevent it from being dulled. With this powerful instrument, the largest hoofs are severed by a moderate compression of the hand. Two well-sharpened knives, which should be kept in a stand or box within reach, are then grasped by the washer and assistant, and with two dexterous strokes to each foot, the side-crust, being free from dirt, and soaked almost as soft as a cucumber, is reduced to the level of the sole. Two expert men will go through these processes in a very short space of time. The closer the paring and clipping the better, if blood be not drawn. An occasional sheep may require clipping again in the fall.

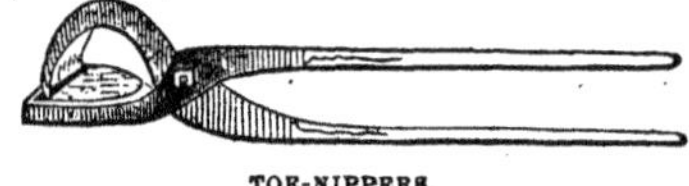
TOE-NIPPERS.

SHEARING.

The time which should elapse between washing and shearing depends altogether on circumstances. From four to six days of bright, warm weather is sufficient; if cold, or rainy, or cloudy, more time must intervene. Sometimes the wool remains in a condition unfit for shearing for a fortnight after washing. The rule to be observed is, that the water should be thoroughly dried out, and the natural oil of the wool should so far exude as to give the wool an unctious feeling, and a lively, glittering look. If it is sheared when dry, like cotton, and before the oil has exuded, it is very difficult to thrust the shears through, the umer is checked, and the wool will not keep so well for long periods. If it is left until it gets too oily, either the manufacturer is cheated, or, what more frequently happens, the owner loses on the price.

FLEECE.

Shearing, in this country, is always done on the threshing-floors of the barns—sometimes upon low platforms, some eighteen or twenty inches high, but more commonly on the floor itself. The place where the sheep remain should be well littered down with straw, and fresh straw thrown on occasionally, to keep the sheep clean while shearing. No chaff or other substance which will stick in the wool should be used for this purpose. The shearing should not commence until the dew, if any, has dried off from the sheep. All loose straws sticking to the wool should be picked off, and whatever dung may adhere to any of the feet brushed off. The floor or tables used should be planed or worn perfectly smooth, so that they will not hold dirt, or catch the wool. They should all be thoroughly cleaned, and, if necessary, washed, preparatory to the process. If there are any sheep in the pen dirty from purging, or other causes, they should first be caught out, to prevent them from contaminating others.

The manner of shearing varies with almost every district; and it is difficult, if not impossible, to give intelligible practical instructions, which would guide an entire novice in skilfully shearing a sheep. Practice is requisite. The following directions are as plain, perhaps, as can be made:

The shearer may place the sheep on that part of the floor assigned to him, resting on its rump, and himself in a posture with his right knee on a cushion, and the back of the animal resting against his left thigh. He grasps the shears about half-way from the point to the bow, resting his thumb along the blades, which gives him better command of the points. He may then commence cutting the wool at the brisket, and, proceeding downward, all upon the sides of the belly to the extremity of the ribs, the external sides of both sides to the edges of the flanks; then back to the brisket, and thence upward, shearing the wool from the breast, front, and both sides of the neck, but not yet the back of it, and also the poll, or forepart, and top of the head. Then "the jacket is opened" of the sheep, and its position, as well as that of the shearer, is changed by the animal's being turned flat upon its side, one knee of the shearer resting on the cushion, and the other gently pressing the fore-quarter of the animal, to prevent any struggling. He then resumes cutting upon the flank and rump, and thence onward to the head. Thus one side is complete. The sheep is then turned on the other side—in doing which great care is requisite to prevent the fleeces being torn—and the shearer proceeds as upon the other, which finishes. He must then take the sheep near to the door through which it is to pass out, and neatly trim the legs, leaving not a solitary lock anywhere as a lodging-place for ticks. It is absolutely necessary for him to remove from his stand to trim, otherwise the useless stuff from the legs becomes intermingled with the fleece-wool. In the use of the shears, the blades should be laid as flat to the skin as possible, the points not lowered too much, nor should more than from one to two inches be cut at a clip, and frequently not so much, depending on the part, and the compactness of the wool.

The wool should be cut off as close as conveniently practicable, and even. It may, indeed, be cut too close, so that the sheep can scarcely avoid sun-scald; but this is very unusual. If the wool is left in ridges, and uneven, it betrays a want of workmanship very distasteful to the really good farmer. Great care should be taken not to cut the wool twice in two, as inexperienced shearers are apt to do, since it is a great damage to the wool. This results from cutting too far from the points of the shears, and suffering them to get too elevated. In such cases, every time the shears are pushed forward, the wool before, cut off by the points, say a quarter or three-eighths of an inch from the hide, is again severed. To keep the fleece entire, which is of great importance to its good appearance when done up, and, therefore, to its salableness, it is very essential that the sheep be held easily for itself, so that it will not struggle violently. No man can hold it still by main strength, and shear it well. The posture of the shearer should be such that the sheep is actually confined to its position, so that it is unable to start up suddenly and tear its fleece; but it should not be confined there by severe pressure or force, or it will be continually kicking and struggling. Clumsy, careless men, therefore, always complain of getting the most troublesome sheep. The neck, for example, may be confined to the floor by placing it between the toe and knee of the leg on which the shearer kneels; but the lazy or brutal shearer who suffers his leg to rest directly on the neck, soon provokes that struggle which the animal is obliged to make to free itself from severe pain, and even, perhaps, to draw its breath.

Good shearers will shear, on the average, twenty-five Merinos per day; but a new beginner should not attempt to exceed from one-third to one-half of that number. It is the last process in the world which should be hurried, as the shearer will, in that case, soon leave more than enough wool on his sheep to pay for his day's wages. Wool ought not to be sheared, and must not be done up with any water in it. If wounds are made, as sometimes happens with unskilful operators, a mixture of tar and grease ought to be applied.

Shearing lambs is, in the Northern climate, at least, an unprofitable practice; since the lamb, at a year old, will give the same amount of wool, and it is thus stripped of its natural protection from cold when it is young and tender, for the mere pittance of the interest on a pound, or a pound and a half of wool for six months, not more than two or three cents, and this all consumed by the expense of shearing. Much the same may be said of the custom, which obtains in some places, of shearing from sheep twice a year. There may be a reason for it, where they receive so little care that a portion are expected to disappear every half year, and the wool to be torn from the backs of the remainder by bushes, thorns, etc., if left for a long period; but when sheep are inclosed, and treated as domestic animals, although there may be less barbarity in shearing them in the fall also, than in the case of the tender lambs, there is no ground for it on the score of utility; since any gain accruing from it cannot pay the additional expense which it occasions.

Cold storms occurring soon after shearing sometimes destroy sheep, in the northern portions of the country, especially the delicate Saxons; forty or fifty of which have, at times, perished out of a single flock, from one night's exposure. Sheep, in such cases, should be housed; or, where this is impracticable, driven into dense forests.

Sun-scald. When they are sheared close in very hot weather, have no shade in their pastures, and especially where they are driven immediately considerable distances, or rapidly, over burning and dusty roads, their backs are sometimes so scorched by the sun that their wool comes off. If let alone, the matter is not a serious one; but the application of refuse lard to the back will hasten the cure, and the starting of the wool.

Ticks. These vermin, when very numerous, greatly annoy and enfeeble the sheep in winter, and should be kept entirely out of the flock. After shearing, the heat and cold, the rubbing and biting of the sheep, soon drive off the tick, and it takes refuge in the wool of the lamb. Let a fortnight elapse after shearing, to allow all to make this change of residence. Then boil refuse tobacco leaves until the decoction is strong enough to kill ticks beyond a peradventure, which may be ascertained by experiment. Five or six pounds of cheap plug tobacco, or an equivalent in stems, and the like, may be made to answer for a hundred lambs.

This decoction is poured into a deep, narrow box, kept for the purpose, which has an inclined shelf on one side, covered with a wooden grate. One man holds the lamb by its hind legs, while another clasps the fore legs in one hand, and shuts the other about the nostrils, to prevent the liquid from entering them, and then the animal is entirely immersed. It is then immediately lifted out, laid on one side upon the grate, and the water squeezed out of its wool, when it is turned over and squeezed on the other side. The grate conducts the fluid back into the box. If the lambs are regularly dipped every year, ticks will never trouble a flock.

MARKING OR BRANDING.

The sheep should be marked soon after shearing, or mistakes may occur. Every sheep-owner should be provided with a marking instrument, which will stamp his initials, or some other distinctive mark, such as a small circle, an oval, a triangle, or a square, at a single stroke, and with uniformity, on the sheep. It is customary to have the mark cut out of a plate of thin iron, with an iron handle terminating in wood; but one made by cutting a type, or raised letter, or character, on the end of a stick of light wood, such as pine or basswood, is found to be better. If the pigment used be thin, and the marker be thrust into it a little too deeply, as often happens, the surplus will not run off from the wood, as it does from a thin sheet of iron, to daub the sides of the sheep, and spoil the appearance of the mark; and, if the pigment be applied hot, the former will not get heated, like the latter, and increase the danger of burning the hide.

Various pigments are used for marking. Many boil tar until it assumes a glazed, hard consistency when cold, and give it a brilliant, black color by stirring in a little lamp-black during the boiling. This is applied when just cold enough not to burn the sheep's hide, and it forms a bright, conspicuous mark all the year round. The manufacturer, however, prefers the substitution of oil and turpentine for tar, as the latter is cleansed out of the wool with some difficulty. It should be boiled in an iron vessel, with high sides, to prevent it from taking fire, on a small furnace or chafing-dish near which it is to be used. When cool enough, forty or fifty sheep can be marked before it gets too stiff. It is then warmed from time to time, as necessary, on the chafing-dish. Paint, made of lampblack, to which a little spirits of turpentine is first added, and then diluted with linseed or lard oil, is also used. The rump is a better place to mark than the side, since it is there about as conspicuous under any circumstances, and more so when the sheep are huddled in a pen, or running away from one. Besides, should any wool be injured by the mark, that on the rump is less valuable than that on the side. Ewes are commonly distinguished from wethers by marking them on different sides of the rump.

Many mark each sheep as it is discharged from the barn by the shearer; but it consumes much less time to do it at a single job, after the shearing is completed; and it is necessary to take the latter course if a hot pigment is used.

MAGGOTS. Rams with horns growing closely to their heads are very liable to have maggots generated under them, particularly if the skin on the surrounding parts becomes broken by fighting; and these, unless removed, soon destroy the animal. Boiled tar, or the marking substance first described, is both remedy and preventive. If it is put under the horns at the time of marking, no trouble will ever arise from this cause.

Sometimes when a sheep scours in warm weather, and clotted dung adheres about the anus, maggots are generated under it, and the sheep perishes miserably. As a preventive, the dung should be removed; as a remedy, the dung and maggots should be removed—the latter by touching them with a little turpentine—and sulphur and grease afterward applied to the excoriated surface.

Maggot-flies sometimes deposit their eggs on the backs of the long, open-woolled English sheep, and the maggots, during the few days before they assume the *pupa* state, so tease and irritate the animal, that fever and death ensue. Tar and turpentine, or butter and sulphur, smeared over the parts, are admirable preventives. The Merino and Saxon are exempt from these attacks.

SHORTENING THE HORNS. A convolution of the horn of a ram sometimes so presses in upon the side of the head or neck that it is necessary to shave or rasp it away on the under side, to prevent ultimately fatal effects. The point of the horn of both ram and ewe both frequently turn in so that they will grow into the flesh, and sometimes into the eye, unless shortened. The toe-nippers will often suffice on the thin extremity of a horn; if not, a fine saw must be used. The marking-time affords the best opportunity for attending to this operation.

SELECTION AND DIVISION.

The necessity of annually weeding the flock, by excluding all its members falling below a certain standard of quality, and the points which should be regarded in fixing that standard, have already been brought to notice in connection with the principles of breeding.

The time of shearing is by far the most favorable period for the flock-master to make his selection. He should be present on the shearing-floor, and inspect the fleece of every sheep as it is gradually taken off; since, if there are faults about it, he will then discover it better than at any other time. A glance will likewise reveal to him every defect in form, previously concealed, wholly, or in part, by the wool, as soon as the newly-shorn sheep is permitted to stand up on its feet. A remarkably choice ewe is frequently retained until she dies of old age; a rather poor nurse or breeder is excluded for the slightest fault, and so on. Whatever animals are to be excluded, may be marked on the shoulder with Venetian red and hog's lard, conveniently applied with a brush or cob. Such of the wethers as have attained their prime, and those ewes that have passed it, should be provided with the best feed, and fitted for the butcher. If they have been properly pushed on grass, they will be in good flesh by the time they are taken from it; and, if not intended for stall-feeding, the sooner they are then disposed of the better.

Those *divisions*, also, in large flocks, which utility demands, are generally made at or soon after shearing. Not more than two hundred sheep should be allowed to run together in the pastures; although the number might, perhaps, be safely increased to three hundred, if the range is extensive.

Wethers and dry ewes to be turned off should be kept separate from the nursing-ewes; and if the flock is large enough to require a third division, it is customary to put the yearling and two-year-old ewes and wethers, and the old, feeble sheep together. It is better, in all cases, to separate the rams from all the other sheep at the time of shearing, and to inclose them in a field which is particularly well-fenced. If they are put even with wethers, they are more quarrelsome; and when cool nights arrive, will worry themselves and waste their flesh in constant efforts to ride the wethers.

The Merino ram, although a quiet animal compared with the common-woolled one, will be tempted to jump, by poor fences, or fences half the time down; and if he is once taught this trick, he becomes very troublesome as the rutting period approaches, unless hoppling, yoking, clogging, or poking is resorted to, either of which causes him to waste his strength, besides being the occasion of frequent accidents.

THE CROOK.

This convenient implement for catching sheep is of the form represented in the cut accompanying, of three-eighths inch round iron, drawn smaller toward the point, which is made safe by a knot. The other end is furnished with a socket, which receives a handle six or eight feet long.

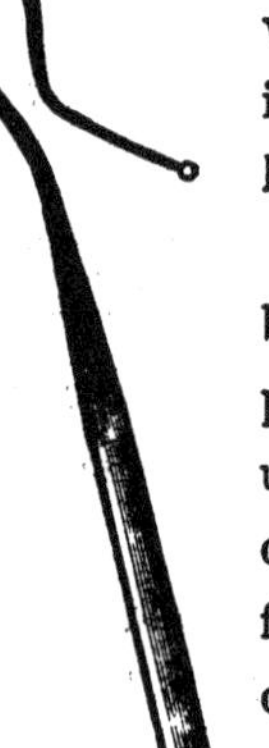

SHEPHERD'S CROOK.

In using it, the hind leg is hooked in from behind the sheep, and it fills up the narrow part beyond that point, while passing along it until it reaches the loop, when the animal is caught by the hook, and when secured, its foot easily slips through the loop. Some caution is required in its use; for, should the animal give a sudden start forward to get away, the moment it feels the crook, the leg will be drawn forcibly through the narrow part, and strike the bone with such violence against the bend of the loop as to cause the animal considerable pain, and even occasion lameness for some days. On first embracing the leg, the crook should be drawn quickly toward the shepherd, so as to bring the bend of the loop against the leg as high up as the hock, before the sheep has time even to break off; and being secure, its struggles will cease the moment the hand seizes the leg.

No shepherd should be without this implement, as it saves much yarding and running, and leads to a prompt examination of every improper or suspicious appearance, and a seasonable application of remedy or preventive, which would often be deferred if the whole flock had to be driven to a distant yard to effect the catching of a single sheep.

Dexterity in its use is speedily acquired by any one; and if a flock are properly tame, any one of its number can be readily caught by it at salting-time, or, generally, at other times, by a person with whom the flock are familiar. It is, however, at the lambing-time, when sheep and lambs require to be so repeatedly caught, that the crook is more particularly serviceable. For this purpose, at that time alone, it will pay for itself ten times over in a single season, in saving time, to say nothing of the advantage of the sheep.

DRIVING AND SLAUGHTERING.

Driving. Mutton can be grown cheaper than any other kind of meat. It is fast becoming better appreciated; and, strange as it may seem, good mutton brings a higher price in our best markets than the same quality does in England. Its substitution in a large measure for pork would contribute materially to the health of the community.

Winter fattening of sheep may often be made very profitable

and deserves greater attention, especially where manure is an object; and the instances are few, indeed, where it is not. In England, it is considered good policy to fatten sheep, if the increase of weight will pay for the oil-cake or grain consumed; the manure being deemed a fair equivalent for the other food—that is, as much straw and turnips as they will eat. Lean sheep there usually command as high a price per pound in the fall as fatted ones in the spring; while, in this country, the latter usually bear a much higher price, which gives the feeder a great advantage.

The difference may be best illustrated by a simple calculation. Suppose a wether of a good mutton breed, weighing eighty pounds in the fall, to cost six cents per pound, amounting to four dollars and eighty cents, and to require twenty pounds of hay per week, or its equivalent in other food, and to gain a pound and a half each week; the gain in weight in four months would be about twenty-five pounds, which, at six cents per pound, would be one dollar and fifty cents, or less than ten dollars per ton for the hay consumed; but if the same sheep could be bought in the fall for three cents per pound, and sold in the spring for six cents, the gain would amount to three dollars and ninety cents, or upwards of twenty dollars per ton for the hay—the manure being the same in either case.

For fattening, it is well to purchase animals as large and thrifty, and in as good condition as can be had at fair prices; and to feed liberally, so as to secure the most rapid increase that can be had without waste of food. The fattening of sheep by the aid of oil-cake, or grain purchased for the purpose, may often be made a cheaper mode of obtaining manure than by the purchase of artificial fertilizers, as guano, super-phosphate of lime, and the like; and it is altogether preferable. It is practised extensively and advantageously abroad, and deserves at least a fair trial among us.

Sheep which are to be driven to market should not begin their journey either when too full or too hungry; in the former state, they are apt to purge while on the road, and in the latter, they will lose strength at once. The sheep selected for market should be those in best condition at the time; and to ascertain this, it is necessary to examine the whole lot, and separate the fattest from the rest, which is best done at about mid-day, before the sheep feed again in the afternoon. The selected ones are placed in a field by themselves, where they remain until the time for starting. If there be rough pasture to give them, they should be allowed to use it, in order to rid themselves of some of the food which might be productive of inconvenience on the journey. If there is no

THE SHEPHERD AND HIS FLOCK.

such pasture, a few cut turnips will answer. All their hoofs should be carefully examined, and every unnecessary appendage removed, though the firm portion of the horn should not be touched. Every clotted piece of wool should also be removed with the shears, and the animals properly marked.

Being thus prepared, they should have feed early in the morning, and be started, in the cold season, about mid-day. Let them walk quietly away; and as the road is new to them, they will go too fast at first, to prevent which, the drover should go before them, and let his dog bring up the rear. In a short time they will assume the proper speed—about one mile an hour. Should the road they travel be a green one, they will proceed nibbling their way onward at the grass along both sides; but if it is a narrow turnpike, the drover will require all his attention in meeting and being passed by various vehicles, to avoid injury to his charge. In this part of their business, drovers generally make too much ado; and the consequence is, that the sheep are driven more from side to side of the road than is requisite. Upon meeting a carriage, it would be much better for the sheep, were the drover to go forward, instead of sending his dog, and point off with his stick the leading sheep to the nearest side of the road; and the rest will follow, as a matter of course, while the dog walks behind the flock and brings up the stragglers. Open gates to fields are sources of great annoyance to drovers, the stock invariably making an endeavor to go through them. On observing an open gate ahead, the drover should send his dog behind him over the fence, to be ready to meet the sheep at the gate. When the sheep incline to rest, they should be allowed to lie down.

When the animals are lodged for the night, a few turnips or a little hay should be furnished to them, if the road-sides are bare. If these are placed near the gate of the field which they occupy, they will be ready to take the road again in the morning. As a precaution against worrying dogs, the drover should go frequently through the flock with a light, retire to rest late, and rise up early in the morning. These precautions are necessary; since, when sheep have once been disturbed by dogs, they will not settle again upon the road. The first day's journey should be a short one, not exceeding four or five miles. The whole journey should be so marked out as that, allowance being made for unforeseen delays, the animals may have one day's rest near the market.

Points of fat sheep. The formation of fat, in a sheep destined to be fattened, commences in the inside, the web of fat which envelopes the intestines being first formed, and a little deposited around the kidneys. After that, fat is seen on the outside; and first upon the end of the rump at the tail-head, continuing to move on along the back, on both sides of the spine, or back-bone, to the bend of the ribs to the neck. Then it is deposited between the muscles, parallel with the cellular tissue. Meanwhile, it is covering the lower round of the ribs descending to the flanks, until the two sides meet under the belly, whence it proceeds to the brisket, or breast, in front, and the sham or cod behind, filling up the inside of the arm-pits and thighs. While all these depositions are proceeding on the outside, the progress in the inside is not checked, but rather increased, by the fattening disposition encouraged by the acquired condition; and, hence, simultaneously, the kidneys become entirely covered, and the space between the intestines and the lumbar region, or loin, gradually filled up by the web and kidney fat.

By this time the cellular spaces around each fibre of muscle are receiving their share; and when fat is deposited there in quantity, it gives to the meat the term *marbled.* These inter-fibrous are the last to receive a deposition of fat; but after this has begun, every other part at the same time receives its due share, the back and kidneys securing the most, so much so that the former literally becomes *nicked*, as it is termed—that is, the fat is felt through the skin to be divided into two portions, from the tail-head along the back to the top of the shoulder; and the tail becoming thick and stiff, the top of the neck broad, the lower part of each side of the neck toward the breasts full, and the hollows between the breast-bone and the inside of the fore legs, and between the cod and the inside of the hind thighs, filled up. When all this has been accomplished, the sheep is said to be *fat*, or *ripe.*

When the body of a fat sheep is entirely overlaid with fat, it is in the most valuable state as mutton. Few sheep, however, lay on fat entirely over their body; one laying the largest proportion on the rump, another on the back; one on the parts adjoining the fore-quarter, another on those of the hind-quarter; and one more on the inside, and another more on the outside. Taking so many parts, and combining any two or more of them together, a considerable variety of condition will be found in any lot of fat sheep, while any one is as ripe in its way as any other.

With these data for guides, the state of a sheep in its progress toward ripeness may be readily detected by handling. A fat sheep, however, is easily known by the eye, from the fullness exhibited by all the external parts of the particular animal. It may exhibit want in some parts when compared with others; but those parts, it may easily be seen, would never become so ripe as the others; and this arises from some constitutional defect in the animal itself; since, if this were so, there is no reason why all the parts should not be alike ripe. The state of a sheep that is obviously not ripe cannot altogether be ascertained by the eye. It must be handled, or subjected to the scrutiny of the hand. Even in so palpable an act as handling, discretion is requisite. A full-looking sheep needs hardly to be handled on the rump; for he would not seem so full, unless fat had first been deposited there. A thin-looking sheep, on the other hand, should be handled on the rump; and if there be no fat there, it is useless to handle the rest of the body, for certainly there will not be so much as to deserve the name of fat. Between these two extremes of condition, every variety exists; and on that account examination by the hand is the rule, and by the eye alone the exception. The hand is, however, much assisted by the eye, whose acuteness detects deficiencies and redundancies at once.

In handling sheep, the points of the fingers are chiefly employed; and the accurate knowledge conveyed by them, through practice, of the exact state of the condition, is truly surprising, and establishes a conviction in the mind that some intimate relation exists between the external and internal state of an animal. Hence originates this practical maxim in judging stock of all kinds—that no animal will appear ripe to the eye, unless as much fat had previously been acquired in the inside as constitutional habit will allow.

The application of this rule is easy. When the rump is

found nicked, on handling, fat is to be found on the back; when the back is found nicked, fat is to be expected on the top of the shoulder and over the ribs; and when the top of the shoulder proves to be nicked, fat may be anticipated on the under side of the belly. To ascertain its existence below, the animal must be *turned up*, as it is termed; that is, the sheep is set upon his rump, with his back down, and his hind feet pointing upward and outward. In this position, it can be seen whether the breast and thighs are filled up. Still, all these alone would not disclose the state of the inside of the sheep, which should, moreover, be looked for in the thickness of the flank; in the fullness of the breast, that is, the space in front from shoulder to shoulder toward the neck; in the stiffness and thickness of the root of the tail; and in the breadth of the back of the neck. All these latter parts, especially with the fullness of the inside of the thighs, indicate a fullness of fat in the inside; that is, largeness of the mass of fat on the kidneys, thickness of net, and thickness of layers between the abdominal muscles. Hence, the whole object of feeding sheep on turnips and the like seems to be to lay fat upon all the bundles of fleshy fibres, called muscles, that are capable of acquiring that substance; for, as to bone and muscle, these increase in weight and extent independently of fat, and fat only increases in their magnitude.

Slaughtering. Sheep are easily slaughtered, and the operation is unattended with cruelty. They require some preparation before being deprived of life, which consists in food being withheld from them for not less than twenty-four hours, according to the season. The reason for fasting sheep before slaughtering is to give time for the paunch and intestines to empty themselves entirely of food, as it is found that, when an animal is killed with a full stomach, the meat is more liable to putrefy, and is not so well flavored; and, as ruminating animals always retain a large quantity of food in their intestines, it is reasonable that they should fast somewhat longer to get rid of it, than animals with single stomachs.

DROVER'S OR BUTCHER'S DOG.

Sheep are placed on their side—sometimes upon a stool, called a killing-stool—to be slaughtered, and, requiring no fastening with cords, are deprived of life by the use of a straight knife through the neck, between its bone and the windpipe, severing the carotid artery and the jugular vein of both sides, from which the blood flows freely out, and the animal soon dies.

The skin, as far as it is covered with wool, is taken off, leaving that on the legs and head, which are covered with hair, the legs being disjointed by the knee. The entrails are removed by an incision along the belly, after the carcass has been hung up by the tendons of the houghs. The net is carefully separated from the viscera, and rolled up by itself; but the kidney fat is not then extracted. The intestines are placed on the inner side of the skin until divided into the *pluck*, containing the heart, lungs, and liver; the bag, containing the stomach; and the *puddings*, consisting of the viscera, or guts. The latter are usually thrown away; though the Scotch, however, clean them and work them up into their favorite *haggis*. The skin is hung over a rope or pole under cover, with the skin-side uppermost, to dry in an airy place.

The carcass should hang twenty-four hours in a clean, cool, airy, dry apartment before it is cut down. It should be cool and dry; for, if warm, the meat will not become firm; and, if damp, a clamminess will cover it, and it will never feel dry, and present a fresh, clean appearance. The carcass is divided in two, by being sawed right down the back-bone. The kidney-fat is then taken out, being only attached to the peritoneum by the cellular membrane, and the kidney is extracted from the *suet*, the name given to sheep-tallow in an independent state.

CUTTING UP. Of the two modes of cutting up a carcass of mutton, the English and the Scotch—of the former, the practice in London being taken as the standard, and of the latter, that of Edinburgh, since more care is exercised in this respect in these two cities—the English is, perhaps, preferable; although the Scotch accomplish the task in a cleanly and workmanlike manner.

The *jigot* is the most handsome and valuable part of the carcass, bringing the highest price, and is either a roasting or a boiling piece. A jigot of Leicester, Cheviot, or South-Down mutton makes a beautiful boiled leg of mutton, which is prized the more the fatter it is—this part of the carcass being never overloaded with fat. The *loin* is almost always roasted, the flap of the flank being skewered up, and it is a juicy piece. Many consider this piece of Leicester mutton, roasted, as too rich; and when warm this is, probably, the case; but a cold roast loin is an excellent summer dish. The *back-rib* is divided into two, and used for very different purposes. The forepart—the neck—is boiled, and makes sweet barely-broth; and the meat, when boiled, or rather the whole simmered for a considerable time beside the fire, eats tenderly. The back-ribs make an excellent roast; indeed, there is not a sweeter or more varied one in the whole carcass, having both ribs and shoulder. The shoulder-blade eats best cold, and the ribs, warm. The ribs make excellent chops, the Leicester and South-Down affording the best. The *breast* is mostly a roasting-piece, consisting of rib and shoulder, and is particularly good when cold. When the piece is large, as of the South-Down or Cheviot, the gristly parts of the ribs may be divided from the true ribs, and helped separately. This piece also boils well; or, when corned for eight days, and served with onion sauce, with mashed turnips in it, there are few more savory dishes at a farmer's table. The *shoulder* is separated before being dressed, and makes an excellent roast for family use, being eaten warm or cold, or carved and dressed as the breast mentioned above. The shoulder is best from a large carcass of South-Down, Cheviot, or Leicester. The *neck-piece* is partly laid bare by the removal of the shoulder, the fore-part being fitted for boiling and making into broth, and the best part for roasting or broiling into chops. On this account, it is a good family piece, and generally preferred to any part of the hind-quarter. Heavy sheep, such as the Leicester, South-Down, and Cheviot, supply the most thrifty neck-piece.

Relative qualities. The different sorts of mutton in common use differ as well in quality as in quantity. The flesh of the *Leicester* is large, though not coarse-grained, of a lively red color, and the cellular tissue between the fibres contains a considerable quantity of fat. When cooked, it is tender and juicy, yielding a red gravy, and having a sweet, rich taste; but the fat is rather too much and too rich for some people's tastes, and can be put aside. It must be allowed that the lean of fat meat is far better than lean meat that has never been fat. *Cheviot* mutton is smaller in the grain, not so bright of color, with less fat, less juice, not so tender and sweet; but the flavor is higher, and the fat not so luscious. The mutton of *South-Downs* is of medium fineness in grain, color pleasant red, fat well intermixed with the meat, juicy, and tenderer than Cheviot. The mutton of rams of any breed is always hard, of disagreeable flavor, and, in autumn, not eatable; that of old ewes is dry, hard, and tasteless; of young ones, well enough flavored, but still rather dry; while wether-mutton is the meat in perfection, according to its kind.

The want of relish, perhaps the distaste, for mutton has served as an obstacle to the extension of sheep husbandry in the United States. The common mistake in the management of mutton among us is, that it is eaten, as a general thing, at exactly the wrong time after it is killed. It should be eaten immediately after being killed, and, if possible, before the meat has time to get cold; or, if not, then it should be kept a week or more—in the ice-house, if the weather require—until the time is just at hand when the fibre passes the state of toughness which it takes on at first, and reaches that incipient or preliminary point in its process toward putrefaction when the fibres begin to give way, and the meat becomes tender.

An opinion likewise generally prevails that mutton does not attain perfection in juiciness and flavor much under five years.

If this be so, that breed of sheep must be very unprofitable which takes five years to attain its full state; and there is no breed of sheep in this country which requires five years to bring it to perfection. This being the case, it must be folly to restrain sheep from coming to perfection until they have reached that age. Lovers of five-year-old mutton do not pretend that this course bestows profit on the farmer, but only insist on its being best at that age. Were this the fact, one of two absurd conditions must exist in this department of agriculture: namely, the keeping a breed of sheep that cannot, or that should not be allowed to, attain to perfection before it is five years old; either of which conditions makes it obvious that mutton cannot be in its *best* state at five years.

The truth is, the idea of mutton of this age being especially excellent, is founded on a prejudice, arising, probably, from this circumstance: before winter food was discovered, which could maintain the condition of stock which had been acquired in summer, sheep lost much of their summer condition in winter, and, of course, an oscillation of condition occurred, year after year, until they attained the age of five years; when their teeth beginning to fail, would cause them to lose their condition the more rapidly. Hence, it was expedient to slaughter them at not exceeding five years of age; and, no doubt, mutton would be high-flavored at that age, that had been exclusively fed on natural pasture and natural hay. Such treatment of sheep cannot, however, be justified on the principles of modern practice; because both reason and taste concur in mutton being at its best whenever sheep attain their perfect state of growth and condition, not their largest and heaviest; and as one breed attains its perfect state at an earlier age than another, its mutton attains its best before another breed attains what is its best state, although its sheep may be older; but taste alone prefers one kind of mutton to another, even when both are in their best state, from some peculiar property. The cry for five-year-old mutton is thus based on very untenable grounds; the truth being that well-fed and fatted mutton is never better than when it gets its full growth in its second year; and the farmer cannot afford to keep it longer, unless the wool would pay for the keep, since we have not the epicures and men of wealth who would pay the butcher the extra price, which he must have, to enable him to pay a remunerating price to the grazier for keeping his sheep two or three years over.

All writers on diet agree in describing mutton as the most valuable of the articles of human food. Pork may be more stimulating, beef perhaps more nutritious, when the digestive powers are strong; but, while there is in mutton sufficient nutriment, there is also that degree of consistency and readiness of assimilation which renders it most congenial to the human stomach, most easy of digestion, and most promotive of human health. Of it, almost alone, can it be said that it is our food in sickness, as well as in health; its broth is the first thing, generally, that an invalid is permitted to taste, the first thing that he relishes, and is a natural preparation for his return to his natural aliment. In the same circumstances, it appears that fresh mutton, broiled or boiled, requires three hours for digestion; fresh mutton, roasted, three and one-fourth hours; and mutton-suet, boiled, four and one-half hours.

Good *ham* may be made of any part of a carcass of mutton, though the leg is preferable; and for this purpose it is cut in the English fashion. It should be rubbed all over with good salt, and a little saltpetre, for ten minutes, and then laid in a dish and covered with a cloth for eight or ten days. After that, it should be slightly rubbed again, for about five minutes, and then hung up in a dry place, say the roof of the kitchen, until used. Wether mutton is used for hams, because it is fat, and it may be cured any time from November to May; but ram-mutton makes the largest and highest-flavored ham, provided it be cured in spring, because it is out of season in autumn.

There is an infallible rule for ascertaining the *age* of mutton by certain marks on the carcass. Observe the color of the breast-bone, when a sheep is dressed—that is, where the breast-bone is separated—which, in a lamb, or before it is one year old, will be quite red; from one to two years old, the upper and lower bone will be changing to white, and a small circle of white will appear round the edges of the other bones, and the middle part of the breast-bone will yet continue red; at three years old, a very small streak of red will be seen in the middle of the four middle bones, and the others will be white; and at four years, all the breast-bone will be of a white or gristly color.

Contributions to manufactures. The products of sheep are not merely useful to man; they provide his luxuries as well. The skin of sheep is made into *leather*, and, when so manufactured with the fleece on, makes comfortable mats for the doors of rooms, and rugs for carriages. For this purpose, the best skins are selected, and such as are covered with the longest and most beautiful fleece. *Tanned sheep-skin* is used in coarse book-binding. *White sheep-skin*, which is not tanned, but so manufactured by a peculiar process, is used as aprons by many classes of workmen, and, in agriculture, as gloves in the harvest; and, when cut into strips, as twine for sewing together the leather coverings and stuffings of horse-collars. *Morocco leather* is made of sheep-skins, as well as of goat-skins, and the bright red color is given to it by cochineal. *Russia leather* is also made of sheep-skins, the peculiar odor of which repels insects from its vicinity, and resists the mould arising from damp, the odor being imparted to it in currying, by the empyreumatic oil of the bark of the birch-tree. Besides soft leather, sheep-skins are made into a fine, flexible, thin substance, known by the name of *parchment;* and, though the skins of all animals might be converted into writing materials, only those of the sheep and the she-goat are used for parchment. The finer quality of the substance, called *vellum*, is made of the skins of kids and dead-born lambs; and for its manufacture the town of Strasburgh has long been celebrated.

Mutton-suet is used in the manufacture of common *candles*, with a proportion of ox-tallow. Minced suet, subjected to the action of high-pressure steam in a digester, at two hundred and fifty or two hundred and sixty degrees of Fahrenheit, becomes so hard as to be sonorous when struck, whiter, and capable, when made into candles, of giving very superior light. *Stearic candles*, the invention of the celebrated Guy Lussac, are manufactured solely from mutton-suet.

Besides the fat, the intestines of sheep are manufactured into various articles of luxury and utility, which pass under the absurd name of *catgut.* All the intestines of sheep are composed of four layers, as in the horse and cattle. The outer, or *peritoneal* one, is formed of that membrane, by which every portion of the belly and its contents is invested, and confined in its natural and proper situation. It is highly smooth and polished, and secretes a watery fluid which contributes to preserve that smoothness, and to prevent all friction and concussion during the different motions of the animal. The second is the *muscular* coat, by means of which the contents of the intestines are gradually propelled from the stomach to the rectum, thence to be expelled when all the useful nutriment is extracted. The muscles, as in all the other intestines, are disposed in two layers, the fibres of the outer coat taking a longitudinal direction, and the inner layer being circular—an arrangement different from that of the muscles of the œsophagus, and in both beautifully adapted to the respective functions of the tube. The *submucous* coat comes next. It is composed of numerous glands, surrounded by cellular tissue, and by which the inner coat is lubricated, so that there may be no obstruction to the passage of the food. The *mucous* coat is the soft villous one lining the intestinal cavity. In its healthy state, it is always covered with mucus; and when the glands beneath are stimulated, as under the action of physic, the quantity of mucus is increased; it becomes of a more watery character; the contents of the intestines are softened and dissolved by it; and by means of the increased action of the muscular coat, which, as well as the mucous one, feels the stimulus of the physic, the fœces are hurried on more rapidly and discharged.

In the manufacture of some sorts of *cords* from the intestines of sheep, the outer peritoneal coat is taken off and manufactured into a thread to sew intestines, and make the cords of rackets and battledores. Future washings cleanse the guts, which are then twisted into different-sized cords for various purposes; some of the best known of which are whip-cords, hatter's cords for bow-strings, clock-maker's cords, bands for spinning-wheels, now almost obsolete, and fiddle and harp-strings. Of the last class, the cords manufactured in Italy are superior in goodness and strength; and the reason assigned is, that the sheep of that country are both smaller and leaner than the breeds most in vogue in England and in this country. The difficulty in manufacturing from other breeds of sheep lies, it seems, in making the treble strings from the fine peritoneal coat, their chief fault being weakness; by reason of which the smaller ones are hardly able to bear the stretch required for the higher notes in concert-pitch, maintaining, at the same time, in their form and construction, that tenuity or smallness of diameter which is required in order to produce a brilliant and clear tone.

The dry and healthful climate, the rolling surface, and the sweet and varied herbage, which generally prevail in the United States, insure perfect health to an originally sound and well-selected flock, unless they are peculiarly exposed to disease. No country is better suited to sheep than most of the Northern and some of the Southern portions of our own. In Europe, and especially in England, where the system of management is, necessarily, in the highest degree artificial, consisting, frequently, in an early and continued forcing of the system, folding on wet, ploughed ground, and the excessive use of that watery food, the Swedish turnip, there are numerous and fatal diseases, a long list of which invariably cumbers the pages of foreign writers on this animal.

The diseases incident to our flocks, on the contrary, may generally be considered as casualties, rather than as inbred, or necessarily arising from the quality of food, or from local causes. It may be safely asserted that, with a dry pasture, well stocked with varied and nutritious grasses; a clear, running stream; sufficient shade and protection against severe storms; a constant supply of salt, tar, and sulphur in summer; good hay, and sometimes roots, with ample shelters in winter —young sheep, originally sound and healthy, will seldom or never become diseased on American soil.

The comparatively few diseases, which it may be necessary here to mention, are arranged in alphabetical order—as in the author's "Cattle and their Diseases"—for convenience of reference, and treated in the simplest manner. Remedies of general application, to be administered often by the unskilful and ignorant, must neither be elaborate nor complicated; and, if expensive, the lives of most sheep would be dearly purchased by their application.

A sheep, which has been reared or purchased at the ordinary price, is the only domestic animal which can die without material loss to its owner. The wool and felt will, in most instances, repay its cost, while the carcass of other animals will be worthless, except for manure. The loss of sheep, from occasional disease, will leave the farmer's pocket in a very different condition from the loss of an equal value in horses or cattle. Humanity, however, alike with interest, dictates the use of such simple remedies, for the removal of suffering and disease, as may be within reach.

ADMINISTERING MEDICINE.

The stomach into which medicines are to be administered is the fourth, or digesting stomach. The comparatively insensible walls of the rumen, or paunch, are but slightly acted upon, except by doses of very improper magnitude. Medicine, to reach the fourth stomach, should be given in a state as nearly approaching fluidity as may be. Even then it may be given in such a manner as to defeat the object in view.

If the animal forcibly gulps fluids down, or if they are given hastily and bodily, they will follow the caul at the base of the gullet with considerable momentum, force asunder the pillars, and enter the rumen; if they are drunk more slowly, or administered gently, they will trickle down the throat, glide over these pillars, and pass on through the maniplus to the true stomach.

BLEEDING.

Bleeding from the ears or tail, as is commonly practised, rarely extracts a quantity of blood sufficient to do any good where bleeding is indicated. To bleed from the eye-vein, the point of a knife is usually inserted near the lower extremity of the pouch below the eye, pressed down, and then a cut made inward toward the middle of the face.

Bleeding from the angular or cheek-vein is recommended, in the lower part of the cheek, at the spot where the root of the fourth tooth is placed, which is the thickest part of the cheek, and is marked on the external surface of the bone of the upper jaw by a tubercle, sufficiently prominent to be very sensible to the finger when the skin of the cheek is touched. This tubercle is a certain index to the angular vein, which is placed below. The shepherd takes the sheep between his legs; his left hand more advanced than his right, which he places under the head, and grasps the under jaw near to the hinder extremity, in order to press the angular vein, which passes in that place, for the purpose of making it swell; he touches the right cheek at the spot nearly equidistant from the eye and mouth, and there finds the tubercle which is to guide him, and also feels the angular vein swelled below this tubercle; he then makes the incision from below upward, half a finger's breadth below the middle of the tubercle. When the vein is no longer pressed upon, the bleeding will commonly cease; if not, a pin may be passed through the lips of the orifice, and a lock of wool tied round them.

For thorough bleeding, the jugular vein is generally to be preferred. The sheep should be firmly held by the head by an assistant, and the body confined between his knees, with its rump against a wall. Some of the wool is then cut away from the middle of the neck over the jugular vein, and a ligature, brought in contact with the neck by opening the wool, is tied around it below the shorn spot near the shoulder. The vein will soon rise. The orifice may be secured, after bleeding, as before described.

The good effects of bleeding depend almost as much on the *rapidity* with which the blood is abstracted, as the *amount* taken. This is especially true in acute diseases. *Either bleed rapidly or do not bleed at all.* The orifice in the vein, therefore, should be of some length, and made lengthwise with the vein. A lancet is by far the best implement; and even a short-pointed penknife is preferable to the bungling gleam.

Bleeding, moreover, should always be resorted to, when it is indicated at all, as nearly as possible to the *commencement* of the malady.

The amount of blood drawn should never be determined by admeasurement, but by constitutional effect—the lowering of the pulse, and indications of weakness In urgent cases—apoplexy, or cerebral inflammation, for example—it would be proper to bleed until the sheep staggers or falls. The quantity of blood in the sheep is less, in comparison, than that in the horse or ox. The blood of the horse constitutes about one-eighteenth part of his weight; and that of the ox at least one-twentieth; while that of the sheep, in ordinary condition, is one-twenty-second. For this reason, more caution should be exercised in bleeding the latter, especially in frequently resorting to it; otherwise, the vital powers will be rapidly and fatally prostrated. Many a sheep has been destroyed by bleeding freely in disorders not requiring it, and in disorders which did require it at the commencement, but of which the inflammatory stage had passed.

FEELING THE. PULSE.

The number of pulsations can be determined by feeling the heart beat on the left side. The femoral artery passes in an oblique direction across the inside of the thigh, and about the middle of the thigh its pulsations and the character of the pulse can be most readily noted. The pulsations per minute, in a healthy adult sheep, are sixty-five in number; though they have been stated at seventy, and even seventy-five.

APOPLEXY.

Soon after the sheep are turned to grass in the spring, one of the best-conditioned sheep in the flock is sometimes suddenly found dead. The symptoms which precede the catastrophe are occasionally noted. The sheep leaps frantically into the air two or three times, dashes itself on the ground, and suddenly rises, and dies in a few minutes.

Where animals in somewhat poor condition are rather forced forward for the purpose of raising their condition, it sometimes happens that they become suddenly blind and motionless; they will not follow their companions; when approached, they run about, knocking their heads against fences, etc.; the head is drawn round toward one side; they fall, grind their teeth, and their mouths are covered with a frothy mucus. Such cases are, unquestionably, referable to a determination of blood to the brain.

Treatment. If the eyes are prominent and fixed, the membranes of the mouth and nose highly florid, the nostrils highly dilated, and the respiration labored and stertorous, the veins of the head turgid, the pulse strong and rather slow, and these symptoms attended by a partial or entire loss of sight and hearing, it is one of those decided cases of apoplexy which require immediate and energetic treatment. Recourse should at once be had to the jugular vein, and the animal bled until an obvious constitutional effect is produced—the pulse lowered, and the rigidity of the muscles relaxed. An aperient should at once follow bleeding; and if the animal is strong and plathoric, a sheep of the size of the Merino would require at least two ounces of Epsom salts, and one of the large mutton sheep, more. If this should fail to open the bowels, half an ounce of the salts should be given, say twice a day.

BRAXY.

This is manifested by uneasiness; loathing of food; frequent drinking; carrying the head down; drawing the back up; swollen belly; feverish symptoms; and avoidance of the flock. It appears mostly in late autumn and spring, and may be induced by exposure to severe storms, plunging in water when hot, and especially by constipation, brought on by feeding on frost-bitten, putrid, or indigestible herbage. Many sheep die on the prairies from this disease, induced by exposure and miserable forage. Entire prevention is secured by warm, dry shelters, and nutritious, dry food.

Treatment. Remedies, to be successful, must be promptly applied. Bleed freely; and to effect this, immersion in a tub of hot water may be necessary, in consequence of the stagnant state of the blood. Then give two ounces of Epsom salts, dissolved in warm water, with a handful of common salt. If this is unsuccessful, give a clyster, made with a pipeful of tobacco, boiled for a few minutes in a pint of water. Administer half; and if this is not effectual, follow with the remainder. Then bed the animal in dry straw, and cover with blankets; assisting the purgatives with warm gruels, followed by laxative provender till well.

BRONCHITIS.

Where sheep are subject to pneumonia, they are liable to bronchitis as well, which is an inflammation of the mucous membrane, which lines the bronchial tubes, or the air-passages of the lungs. The *symptoms* are those of an ordinary cold, but attended with more fever, and a tenderness of the throat and belly when pressed upon.

Treatment. Administer salt in doses of from one and a half to two ounces, with six or eight ounces of lime-water, given in some other part of the day.

CATARRH.

This is an inflammation of the mucous membrane, which lines the nasal passages, and it sometimes extends to the larynx and pharynx. In the first instance—where the lining of the nasal passages is alone and not very violently affected—it is merely accompanied by an increased discharge of mucus, and is rarely attended with much danger. In this form, it is usually termed *snuffles*; and high-bred English mutton-sheep, in this country, are apt to manifest more or less of it, after every sudden change of weather. When the inflammation extends to the mucous lining of the larynx and the pharynx, some degree of fever usually supervenes, accompanied by cough, and some loss of appetite. At this point, bleeding and purging are serviceable.

Catarrh rarely attacks the American fine-woolled sheep with sufficient violence in summer to require the application of remedies. Depletion, in catarrh, in our severe winter months, however, rapidly produces that fatal prostration, from which it is almost impossible to bring the sheep back, without bestowing an amount of time and care upon it, costing far more than the worth of an ordinary animal.

The best course is to *prevent* the disease by judicious precaution. With that amount of attention which every prudent farmer should bestow on his sheep, the American Merino is but little subject to it. Good, comfortable, and well-ventilated shelters, constantly accessible to the sheep in winter, with a sufficiency of food regularly administered, are usually a sufficient safeguard.

MALIGNANT EPIZOÖTIC CATARRH.

Essentially differing, in type and virulence, from the preceding, is an epidemic, or, more properly speaking, an epizoötic

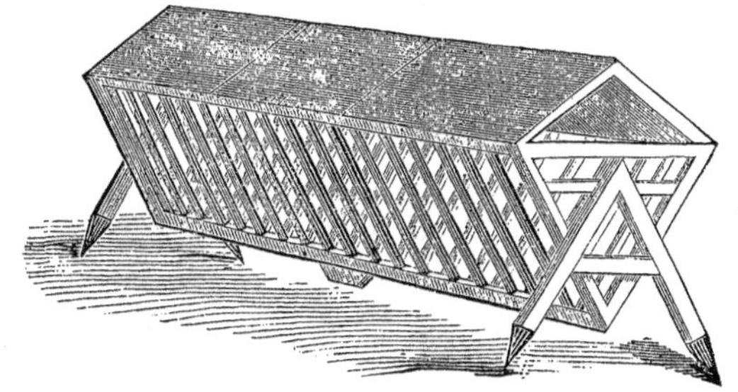

AN ENGLISH RACK FOR FEEDING SHEEP.

malady, which, as often as once in every eight or ten years, sweeps over extended sections of the Northern States, destroying more sheep than all other diseases combined. It commonly makes its appearance in winters characterized by rapid and violent changes of temperature, which are spoken of by the farmers as "bad winters" for sheep. The disease is sometimes termed the "distemper," and also, but erroneously, "grub in the head." The winter of 1846–7 proved peculiarly destructive to sheep in New York, and some of the adjoining States; some owners losing one-half, others three-quarters, and a few seven-eighths, of their flocks. One person lost five hundred out of eight hundred; another, nine hundred out of a thousand. These severe losses, however, mainly fell on the holders of the delicate Saxons, and perhaps, generally, on those possessing not the best accommodations, or the greatest degree of energy and skill.

Symptoms. The primary and main disease, in such instances, is a species of catarrh; differing, however, from ordinary catarrh in its diagnosis, and in the extent of the lesions accompanying both the primary and the symptomatic diseases. The animals affected do not, necessarily, at first show any signs of violent colds, as coughing, sneezing, or labored respiration; the only indications of catarrh noticed, oftentimes, being a nasal discharge. Animals having this discharge appear dull and drooping; their eyes run a little, and are partially closed; the caruncle and lids look pale; their movements are languid, and there is an indisposition to eat; the pulse is nearly natural, though at times somewhat too languid. In a few days these symptoms are evidently aggravated; there is rapid emaciation, accompanied with debility; the countenance is exceedingly dull and drooping; the eye is kept more than half closed; the caruncle, lids, etc., are almost bloodless; a gummy, yellow secretion about the eye; thick, glutinous mucus adhering in and about the nostrils; appetite feeble; pulse languid; and muscular energy greatly prostrated. They rapidly grow weaker, stumble, and fall as they walk, and soon become unable to rise; the appetite grows feebler; the mucus at the nose is, in some instances, tinged with dark, grumous blood; the respiration becomes oppressed; and the animals die within a day or two after they become unable to rise. Upon a *post-mortem* examination, the mucous membrane lining the whole nasal cavity is found highly congested and thickened throughout its entire extent, accompanied with the most intense inflammation; slight ulcers are found on the membranous lining, at the junction of the cellular ethmoid bones with the cribriform plate, in the ethmoidal cells; and the inflammation extends to the mucous membrane of the pharynx, and some inches, from two to four, of the upper portion of the œsophagus.

No sheep, affected with this disease, recovers after emaciation and debility have proceeded to any great extent. In the generality of instances, the time, from the first observed symptoms until death, varies from ten to fifteen days; although death, in some cases, results more speedily.

Treatment. Nothing has been found so serviceable as mercury, which, from its action on the entire secretory system, powerfully tends to relieve the congested membranes of the head. Dissolve one grain of bi-chloride of mercury—corrosive sublimate—in two ounces of water; and give one-half ounce of the water, or one-eighth of a grain of corrosive sublimate, daily, in two doses. To stimulate and open the bowels, give, also, rhubarb in a decoction, the equivalent of ten or fifteen grains at a dose, accompanied with the ordinary carminative and stomachic adjuvants, ginger and gentian in infusion.

COLIC.

Sheep are occasionally seen, particularly in the winter, lying down and rising every moment or two, and constantly stretching their fore and hind legs so far apart that their bellies almost touch the ground. They appear to be in much pain, refuse all food, and not unfrequently die, unless relieved.

This disease, popularly known as the "stretches," is erroneously attributed to an involution of one part of the intestine within another; it being, in reality, a species of flatulent colic, induced by costiveness.

Treatment. Half an ounce of Epsom salts, a drachm of Jamaica ginger, and sixty drops of essence of peppermint. The salts alone, however, will effect the cure; as will, also, an equivalent dose of linseed oil, or even hog's lard.

COSTIVENESS.

This difficulty is removed by giving two table-spoonfuls of castor oil every twelve hours, till the trouble ceases; or give one ounce of Epsom salts. This may be assisted by an injection of warm weak suds and molasses.

DIARRHŒA.

Common diarrhœa—purging, or scours—manifests itself simply by the copiousness and fluidity of the alvine evacuations. It is generally owing to improper food, as bad hay, or noxious weeds; to a sudden change, as from dry food to fresh grass; to an excess, as from overloading the stomach; and sometimes to cold and wet. It is important to clearly distinguish this disease from dysentery. In diarrhœa, there is no apparent general fever; the appetite remains good; the stools are thin and watery, but unaccompanied with slime, or mucus, and blood; odor of the fæces is far less offensive than in dysentery; and the general condition of the animal is but little changed. When it is light, and not of long continuance, no remedy is called for, since it is a healthful provision of Nature for the more rapid expulsion of some offending matter in the system, which, if retained, might lead to disease.

Treatment. Confinement to dry food for a day or two, and a gradual return to it, often suffices, in the case of grown sheep. With lambs, especially if attacked in the fall, the disease is more serious. If the purging is severe, and especially if any mucus is observed with the fæces, the feculent matter should be removed from the bowels by a gentle cathartic; half a drachm of rhubarb, or an ounce of linseed oil, or half an ounce of Epsom salts to a lamb. This should be followed by an astringent; and, in nine cases out of ten, the latter will serve in the first instance. Give one quarter of an ounce of prepared chalk in half a pint of tepid milk, once a day for two or three days; at the end of which, and frequently after the first dose, the purging will have ordinarily abated, or entirely ceased.

"Sheep's cordial" is also a safe and excellent remedy—in severe cases, better than simple chalk and milk. Take of prepared chalk, one ounce; powdered catechu, half an ounce; powdered Jamaica ginger, two drachms; and powdered opium, half a drachm; mix with half a pint of peppermint water; give two or three table-spoonfuls morning and night to a grown sheep, and half that quantity to a lamb.

DISEASE OF THE BIFLEX CANAL.

From the introduction of foreign bodies into the biflex canal, or from other causes, it occasionally becomes the seat of inflammation. This canal is a small orifice, opening externally on the point of each pastern, immediately above the cleft between the toes. It bifurcates within, a tube passing down on each side of the inner face of the pastern, winding round and ending in a *cul de sac*. Inflammation of this canal causes an enlargement and redness of the pastern, particularly about the external orifice of the canal. The toes are thrown wide apart by the tumor. It rarely attacks more than one foot, and should not be allowed to proceed to the point of ulceration which it will do, if neglected. There is none of that soreness and disorganization between the back part of the toes, and none of that peculiar fetor which distinguishes the hoof-ail, with which disease it is sometimes confounded.

Treatment. Scarify the coronet, making one or two deeper incisions in the principal swelling around the mouth of the canal; and cover the foot with tar.

DYSENTERY.

This is occasioned by an inflammation of the mucous or inner coat of the larger intestines, causing a preternatural increase in their secretions, and a morbid alteration in their character. It is frequently consequent on that form of diarrhœa, which is caused by an inflammation of the mucous coat of the smaller intestines. The inflammation extends throughout the whole alimentary canal, increases in virulence, and becomes dysentery, a disease frequently dangerous and obstinate in its character, but, fortunately, not common among sheep, generally, in the United States. Its diagnosis differs from that of diarrhœa, in several readily observed particulars. There is evident fever; the appetite is capricious, commonly very feeble; the stools are as thin as in diarrhœa, or even thinner, but much more adhesive, in consequence of the presence of large quantities of mucus. As the erosion of the intestines advances, the fæces are tinged with blood; their odor is intolerably offensive; and the animal rapidly wastes away, the course of the disease extending from a few days to several weeks.

Treatment. Moderate bleeding should be resorted to, in the first or inflammatory shape, or whenever decided febrile symptoms are found to be present. Two doses of physic having been administered, astringents are serviceable. The "sheep's cordial," already described, is as good as any; and to this, tonics may soon begin to be added; an additional quantity of ginger may enter into the composition of the cordial, and gentian powder will be an useful auxiliary. With this, as an excellent stimulus to cause the sphincter of the anus to contract, and also the mouths of the innumerable secretory and exhalent vessels opening on the inner surface of the intestines, a half grain of strychnine may be combined. Smaller doses should be given for three or four days.

FLIES.

The proper treatment, upon the appearance of flies or maggots, has already been detailed under the head of "FEEDING AND MANAGEMENT," to which the reader is referred.

FOULS.

Sheep are much less subject to this disease than cattle are; but encounter it, if kept in wet, filthy yards, or on moist, poachy ground. It is an irritation of the integument in the cleft of the foot, slightly resembling incipient hoof-ail, and producing lameness. It occasions, however, no serious structural disorganization, disappears without treatment, is not

14

contagious, and appears in the wet weather of spring and fall, instead of in the dry, hot period of summer, when the hoof-ail rages most. A little solution of blue vitriol, or a little spirits of turpentine—either followed by a coating of warm tar—promptly cures it.

For foul noses, dip a small swab in tar, then roll it in salt; put some on the nose, and compel the sheep to swallow a small quantity.

FRACTURES.

If there be no wound of the soft parts, the bone simply being broken, the treatment is extremely easy. Apply a piece of wet leather, taking care to ease the limb when swelling supervenes. When the swelling is considerable, and fever present, the best course is to open a vein of the head or neck, allowing a quantity of blood to escape, proportioned to the size and condition of the animal, and the urgency of the symptoms. Purgatives in such cases should never be neglected. Epsom salts, in ounce doses, given either as a gruel or a drench, will be found to answer the purpose well. If the broken bones are kept steady, the cure will be complete in from three to four weeks, the process of reunion always proceeding faster in a young than in an old sheep. Should the soft parts be injured to any extent, or the ends of the bone protrude, recovery is very uncertain; and it will become a question whether it would not be better to convert the animal at once into mutton.

GARGET.

This is an inflammation of the udder, sometimes known as "caked bag," with or without general inflammation. Where it is simply an inflammation of the udder, it is usually caused by too great an accumulation of milk in the latter prior to lambing, or in consequence of the death of the lamb.

Treatment. Drawing the milk partly from the bag, so that the hungry lamb will butt and work at it an unusual time in pursuit of its food, and bathing it a few times in *cold* water, usually suffices. If the lamb is dead, the milk should be drawn a few times, at increasing intervals, washing the udder for some time in cold water at each milking. In cases of obdurate induration, the udder should be anointed with iodine ointment. If there is general fever in the system, an ounce of Epsom salts may be given. If suppuration forms, the part affected should be opened with the lancet.

GOITRE.

The "swelled neck" in lambs is, like the goitre, or bronchocele, an enlargement of the thyroid glands, and is strikingly analogous to that disease, if not identical with it. It is congenital. The glands at birth are from the size of a pigeon's egg to that of a hen's egg, though more elongated and flattened than an egg in their form. The lamb is exceedingly feeble, and often perishes almost without an effort to suck. Many even make no effort to rise, and die as soon as they are dropped. It is rare, indeed, that one lives.

A considerable number of lambs annually perish from this disease, which does not appear to be an epizoötic, though it is more prevalent in some seasons than in others. It does not seem to depend upon the water, or any other natural circumstances of a region, as goitre is generally supposed to, since it may not prevail in the same flock, or on the same farm, once in ten years; nor can it be readily traced to any particular kind of food. When it does appear, however, its attacks are rarely isolated; from which circumstance some have inferred that it is induced by some local or elimentary cause. Losses from this disease have ranged from ten per cent. to twenty, thirty, and even fifty per cent. of the whole number of lambs. Possibly, high condition in the ewes may be one of the inducing causes.

Treatment. None is known which will reach the case. Should one having the disease chance to live, it would scarcely be worth while to attempt reducing the entanglement of the glands Perhaps keeping the breeding-ewes uniformly in fair, plump, but not *high* condition, would be as effectual a preventive as any.

GRUB IN THE HEAD.

What is popularly known as the "grub" is the larva of the *œstrus oris*, or gad-fly of the sheep. It is composed of five rings; is tiger-colored on the back and belly, sprinkled with spots and patches of brown; its wings are striped.

The sheep gad-fly is led by instinct to deposit its eggs within the nostrils of the sheep. Its attempts to do this—most common in July, August, and September—are always indicated by the sheep, which collect in close clumps, with their heads inward, and their noses thrust close to the ground, and into it, if any loose dirt or sand is within reach. If the fly succeeds in depositing its egg, the latter is immediately hatched by the warmth and moisture of the part, and the young grubs, or larvæ, crawl up the nose, finding their devious way to the sinuses, where, by means of their tentaculæ, or feelers, they attach themselves to the mucous membrane lining those cavities. During the ascent of the larvæ, the sheep stamps, tosses its head violently, and often dashes away from its companions wildly over the field. The larvæ remain on the sinuses, feeding on the mucus secreted by the membrane, and apparently creating no further annoyance, until ready to assume their *pupa* form in the succeeding spring.

Having remained in the sinuses during the fall and winter, they abandon them as the warm weather approaches in the latter part of spring. They crawl down the nose, creating even greater irritation and excitement than when they originally ascended, drop on the ground, and rapidly burrow into it. In a few hours, the skin of the larvæ has contracted, become of a dark-brown color, and it has assumed the form of chrysalis. This fly never eats; the male, after impregnating two or three females, dies; and the latter, having deposited their *ova* in the nostrils of the sheep, also soon perish.

The larvæ in the heads of sheep may, and probably do, add to the irritation of those inflammatory diseases, such as catarrh, which attack the membranous lining of the nasal cavities; and they are a powerful source of momentary irritation in the first instance, when ascending to, and descending from, their lodging-place in the head. But in the interval between these events, extending over a period of several months, not a movement of the sheep indicates the least annoyance at their presence. They are, moreover, found in the heads of nearly all sheep, the healthy as well as the diseased, at the proper season.

Treatment. Though the presence of the grub constitutes no disease, some think it well to diminish their number by all convenient means. One simple way of effecting this is, by turning up with a plough a furrow of earth in the sheep-pasture, into which the sheep will thrust their noses on the approach of the *œstrus*, and thus many of them escape its attacks. Some farmers smear the noses of their sheep occasionally with tar, the odor of which is believed to repel the fly. Another plan, deemed efficacious in dislodging the larvæ from the sinuses, is as follows: Take half a pound of good Scotch snuff, and two quarts of boiling water; stir, and let it stand till cold. Inject about a table-spoonful of this liquid and sediment up each nostril, with a syringe; repeat this three or four times, at intervals, from the middle of October till January. The efficacy of the snuff will be increased by adding half an ounce of asafœtida, pounded in a little water. The effects on the sheep are immediate prostration and apparent death; but they will soon recover. A decoction of tobacco affords a substitute for snuff; and some recommend blowing tobacco smoke through the tail of a pipe into each nostril.

HOOF-AIL.

The first symptom of this troublesome malady, known, likewise, as foot-ail, which is ordinarily noticed, is a lameness of one or both of the forefeet. On daily examining, however, the feet of a flock which have the disease among them, it will readily be seen that the lesions manifest themselves for several days before they are followed with lameness.

The horny covering of the sheep's foot extends up, gradually thinning out, some way between the toes and divisions of the hoof, and above these horny walls the cleft is lined with skin. When the points of the toes are spread apart, this skin is shown in front, covered with short, soft hair. The back part of the toes, or the heels, can be separated only to a little distance, and the skin in the cleft above them is naked. In a healthy foot, the skin throughout the whole cleft is as firm, dry, and uneroded as on any other part of the animal.

The first symptom of hoof-ail is a slight erosion, accompanied with inflammation and heat of the naked skin in the *back parts* of the clefts, immediately above the heels. The skin assumes a macerated appearance, and is kept moist by the presence of a sanious discharge from the ulcerated surface. As the inflammation extends, the friction of the parts causes pain, and the sheep limps. At this stage, the foot, *externally*, in a great majority of cases, exhibits not the least trace of disease, with the exception of a slight redness, and sometimes the appearance of a small sore at the upper edge of the cleft, when viewed from behind.

The ulceration of the surface rapidly extends. The thin upper edges of the inner walls of the hoof are disorganized, and an ulceration is established between the hoof and the fleshy sole. A purulent fetid matter is discharged from the cavity. The extent of the separation increases daily, and the ulcers also form sinuses deep into the fleshy sole. The bottom of the hoof disappears, eaten away by the acrid matter, and the outer walls, entirely separated from the flesh, hang only by their attachments at the coronet. The whole fleshy sole is now entirely disorganized, and the entire foot is a mass of black, putrid ulceration; or, as more commonly happens, the fly has struck it, and a dense mass of writhing maggots cover the surface, and burrow in every cavity.

The forefeet are generally first attacked; and, most usually but one of them. The animal at first manifests but little constitutional disturbance, and eats as usual. By the time that any considerable disorganization of the structures has taken place in the first foot, and sometimes sooner, the other forefoot is attacked. That becoming as lame as the first, the miserable animal seeks its food on its knees; and, if forced to rise, its strange, hobbling gait betrays the intense agony occasioned by bringing its feet in contact with the ground. There is a bare spot under the brisket, of the size of a man's hand, which looks red and inflamed. There is a degree of general fever, and the appetite is dull. The animal rapidly loses condition. The appearance of the maggot soon closes the scene. Where the rotten foot is brought in contact with the side, in lying down, the filthy, ulcerous matter adheres to, and saturates the short wool—it being but a month and a half, or two months, after shearing—and maggots are either carried there by the foot, or they are soon generated there. A black crust is speedily formed round the spot, which is the decomposition of the surrounding structures; and innumerable maggots are at work below, burrowing into the integuments and muscles, and eating up the wretched animal alive. The black, festering mass rapidly spreads, and the poor sufferer perishes, apparently in tortures the most excruciating.

Sometimes but one forefoot is attacked, and subsequently one or both hind ones. There is no uniformity in this particular; and it is a singular fact that, when two or even three of the feet are dreadfully diseased, the fourth may be entirely sound. So, also, one foot may be cured, while every other one is laboring under the malady. The highly offensive odor of the ulcerated feet is so peculiar that it is strictly characteristic of the disease, and would reveal its character, to one familiar with it, in the darkest night.

Hoof-ail is probably propagated in this country exclusively by inoculation—the contact of the matter of a diseased foot with the integuments lining the bifurcation of a healthy foot. That it is propagated in some of the ways classed under the ordinary designation of *contagion*, is certain. That it may be propagated by inoculation, has been established by experiment. The matter of diseased feet has been placed on the skin lining the cleft of a healthy foot under a variety of circumstances—sometimes when that skin is in its ordinary and natural state, sometimes after a very slight scorification, sometimes when macerated by moisture; and under each of these circumstances the disease has been communicated. The same inference may be drawn, also, from the manner in which the disease attacks flocks. The whole, or any considerable number, though sometimes rapidly, are never *simultaneously* attacked, as would be expected, among animals so gregarious, if the disease could be transmitted by simple contact, inhaling the breath, or other effluvium.

The matter of diseased feet is left on grass, straw, and other substances, and thus is brought in contact with the inner surfaces of healthy feet. Sheep, therefore, contract the disease from being driven over the pastures, yarded on the straw, etc., where diseased sheep have been, perhaps even days, before. The matter would probably continue to inoculate, until dried up by the air and heat, or washed away by the rains. The stiff, upright stems of closely mown grass, as on meadows, are almost as well calculated to receive the matter of diseased feet, and deposit it in the clefts of healthy ones, as any means which could be artificially devised. It is not entirely safe to drive healthy sheep over roads, and especially into washing-yards, or sheep houses, where diseased sheep have been, until rain has fallen, or sufficient time has elapsed for the matter to dry up. On the moist bottom of a washing-yard, and particularly in houses or sheds, kept from sun and wind and rain, this matter might be preserved for some time in a condition to inoculate.

When the disease has been well kept under during the first season of its attack, but not entirely eradicated, it will almost or entirely disappear as cold weather approaches, and it does not manifest itself until the warm weather of the succeeding summer. It then assumes a mitigated form; the sheep are not rapidly and simultaneously attacked; there seems to be less inflammatory action constitutionally, and in the diseased parts; the course of the disease is less malignant and more tardy, and it more readily yields to treatment. If well kept under the second summer, it is still milder the third. A sheep will occasionally be seen to limp; but its condition will scarcely be affected, and dangerous symptoms will rarely supervene. One or two applications made during the summer, in a manner presently to be described, will suffice to keep the disease under. At this point, a little vigor in the treatment will rapidly extinguish the disease.

Treatment. The preparation of the foot, where any separate individual treatment is resolved upon, is always necessary, at least in bad cases. Sheep should be yarded for the operation immediately after a rain, if practicable, as the hoofs can then be readily cut. In a dry time, and after a night which left no dew upon the grass, their hoofs are almost as tough as horn. They must be driven through no mud, or soft dung, on their way to the yard, which would double the labor of cleaning their feet. The yard should be small, so that they can be easily caught, and it must be kept well littered down, to prevent their filling their feet with their own excrement. If the straw is wetted, their hoofs will not, of course, dry and harden as rapidly as in dry straw. If the yard could be built over a shallow, gravelly-bottomed brook, it would be an admirable arrangement; for this purpose, a portion of any little brook might be prepared, by planking the bottom, and widening it, if desirable. By such means the hoofs would be kept so soft that the greatest and most unpleasant part of the labor, as ordinarily performed, would be in a great measure saved, and they would be kept free from that dung which, by any other arrangement, will, more or less, get into their clefts.

The principal operator seats himself on a chair, having within his reach a couple of good knives, a whetstone, the powerful toe-nippers already described, a bucket of water with a couple of linen rags in it, together with such medicines as may be deemed necessary. The assistant catches a sheep and lays it partly on its back and rump, between the legs of the foreman, the head coming up about to his middle. The assistant then kneels on some straw, or seats himself on a low stool at the hinder extremity of the sheep. If the hoofs are long, and especially if they are dry and tough, the assistant presents each foot to the operator who shortens the hoof with the toe-nippers. If there is any filth between the toes, each man takes his rag from the bucket of water, and draws it between the toes, and rinses it, until the filth is removed Each then takes a knife, and the process of paring away the horn commences, *upon the effectual performance of which* all else depends. A glance at the foot will show whether it is the seat of the diseased action. The least experience cannot fail in properly settling this question. An experienced finger, even, placed upon the back of the pastern close above the heel, will at once detect the local inflammation, in the dark, *by its heat.*

If the disease is in the first stage—that is, if there are merely erosion and ulceration of the cuticle and flesh in the cleft *above* the walls of the hoof—no paring is necessary. But if ulceration has established itself between the hoof and the fleshy sole, *the ulcerated parts*, however extensive, *must be entirely stripped of their horny covering*, no matter what amount of time and care it may require. It is better not to wound the sole so as to cause it to bleed freely, as the running blood will wash off the subsequent application; but no fear of wounding the sole must prevent a full compliance with the rule laid down above. At the worst, the blood will stop flowing after a little while, during which time no application needs to be made to the foot.

If the foot is in the third stage—a mass of rottenness, and filled with maggots—pour, in the first place, a little spirits of turpentine—a bottle of which, with a quill through the cork, should be always ready—on the maggots, and most of them will immediately decamp, and the others can be removed with a probe or small stick. Then *remove every particle of loose horn*, though it should take the entire hoof, as it generally will in such cases. The foot should next be cleansed with a solution of chloride of lime, in the proportion of one pound of chloride to one gallon of water. If this is not at hand, plunging the foot repeatedly in hot water, just short of scalding hot, will answer every purpose. The great object is *to clean the foot thoroughly*. If there is any considerable "proud flesh," it should be removed with a pair of scissors, or by the actual cautery—hot iron.

The following are some of the most popular remedies: Take two ounces of blue vitriol and two ounces of verdigris, to a junk-bottle of wine; or spirits of turpentine, tar, and verdigris in equal parts; or three quarts of alcohol, one pint of spirits of turpentine, one pint of strong vinegar, one pound of blue vitriol, one pound of copperas, one and a half pounds of verdigris, one pound of alum, and one pound of saltpetre, pounded fine; mix in a close bottle, shake every day, and let it stand six or eight days before using; also mix two pounds of honey and two quarts of tar, which must be applied after the preceding compound. Or apply diluted aquafortis—nitric acid—with a feather to the ulcerated surface; or diluted oil of vitriol—sulphuric acid—in the same way; or the same of muriatic acid; or dip the foot in tar nearly at the boiling point.

In the first and second stages of the disease, before the ulcers have formed sinuses into the sole, and wholly or partly destroyed its structure, the best application is a saturated solution of blue vitriol—sulphate of copper. In the third stage, when the foot is a festering mass of corruption, after it has been cleansed as already directed, it requires some strong caustic to remove the unhealthy granulations—the dead muscular structures—and to restore healthy action. Lunar caustic, which is preferable to any other application, is too expensive; chloride of antimony is excellent, but frequently unattainable in the country drug-stores; and muriatic acid, or even nitric or sulphuric acids, may be used instead. The diseased surface is touched with the caustic, applied with a swab, formed by fastening a little tow on the end of a stick, until the objects above pointed out are attained. The foot is then treated with the solution of blue vitriol, and subsequently coated over with tar which has been boiled, and is properly cooled, for the purpose of protecting the raw wound from dirt, flies, etc. Sheep in this stage of the disease should certainly be separated from the main flock, and looked to as often as once in three days. With this degree of attention, their cure will be rapid, and the obliterated structures of the foot will be restored with astonishing rapidity.

The common method of using the solution of blue vitriol is to pour it from a bottle with a quill in the cork, into the foot, when the animal lies on its back between the operators, as already described. In this way a few cents' worth of vitriol will answer for a large number of sheep. The method is, however, imperfect; since, without extraordinary care, there will almost always be some slight ulcerations not uncovered by the knife, which the solution will not reach, the passages to them being devious, and perhaps nearly or quite closed. The disease will thus be only temporarily suppressed, not cured.

A flock of sheep which were in the second season of the disease, had been but little looked to during the summer, and as cold weather set in, many of them became considerably lame, and some of them quite so. Their feet were thoroughly pared; and into a large washing-tub, in which two sheep could conveniently stand, a saturated solution of blue vitriol and water, *as hot as could be endured by the hand even for a moment*, was poured. The liquid was about four inches deep on the bottom of the tub, and was kept at that depth by frequent additions of hot solution. As soon as a sheep's feet were pared, it was placed in the tub, and held there by the neck. A second one was then prepared, and placed beside it; when the third was ready, the first was taken out; and so on. Two sheep were thus constantly in the tub, each remaining some five minutes. The cure was perfect; there was not a lame sheep in the flock during the winter or the next summer. The hot liquid penetrated to every cavity of the foot; and doubtless had a far more decisive effect, even on the uncovered ulcers, than would have been produced by merely wetting them. The expense attending the operation was about *four cents* per sheep. Three such applications, at intervals of a week, would effectually cure the disease, since every new case would thus be arrested and cured before it would have time to inoculate others. It would, undoubtedly, accomplish this at any time of year, and even during the first and most malignant prevalence of the contagion, *provided the paring was sufficiently thorough*. The second and third parings would be a mere trifle; and the liquid left at the first and second applications could again be used. Thus sheep could be cured at about twelve cents per head, which is much cheaper, in the long run, than any ordinary temporizing method, where the cost of a few pounds of blue vitriol is counted, but not the time consumed; and the disease is thus kept lingering in the flock for years.

Some Northern farmers drive their sheep over dusty roads as a remedy for this disease; and in cases of ordinary virulence, especially where the disease is chronic, it seems to dry up the ulcers, and keep the malady under. Sheep are also sometimes cured by keeping them on a dry surface, and driving them over a barn-floor daily, which is well covered with quicklime. It may sometimes, and under peculiar circumstances, be cured by dryness, and repeated washing with soap-suds.

Many farmers select rainy weather as the time for doctoring their sheep. Their feet are then soft, and it is therefore on all accounts good economy, when the feet are to be pared, and each separately treated, *provided* they can be kept in sheep-houses, or under shelters of any kind, until the rain is over, and the grass again dry. If immediately let out in wet grass of any length, the vitriol or other application is measurably washed away. This is avoided by many, by dipping the feet in more tar—an admirable plan under such circumstances.

A flock of sheep which have been cured of the hoof-ail, is considered more valuable than one which has never had it. They are far less liable to contract the disease from any casual exposure; and its ravages are far less violent and general among them.

This ailment should not be confounded with a temporary soreness, or inflammation of the hoof, occasioned by the irritation from the long, rough grasses which abound in low situations, which is removed with the cause; or, if it continues, white paint or tar may be applied, after a thorough washing.

HOOVE.

This is not common, to any dangerous degree, among sheep; but, if turned upon clover when their stomachs are empty, it will sometimes ensue.

Hoove is a distension of the paunch by gas extricated from the fermentation of its vegetable contents, and evolved more rapidly, or in larger quantities, than can be neutralized by the natural alkaline secretions of the stomach. When the distention is great, the blood is prevented from circulating in the vessels of the rumen, and is determined to the head. The diaphragm is mechanically obstructed from making its ordinary contractions, and respiration, therefore, becomes difficult and imperfect. Death, in such cases, soon supervenes.

Treatment. In ordinary cases, gentle but prolonged driving will effect a cure. When the animal appears swelled almost to bursting, and is disinclined to move, it is better to open the paunch at once. At the most protruberant point of the swelling, on the left side, a little below the hip bone, plunge a trocher or knife, sharp at the point and dull on the edge, into the stomach. The gas will rapidly escape, carrying with it some of the liquid and solid contents of the stomach. If no measures are taken to prevent it, the peristaltic motion, as well as the collapse of the stomach, will soon cause the orifices through the abdomen and paunch not to coincide, and thus portions of the contents of the former will escape into the cavity of the latter.

However perfect the cure of hoove, these substances in the belly will ultimately produce fatal irritation. To prevent this, a canula, or little tube, should be inserted through both orifices as soon as the puncture is made. Where the case is not imminent, alkalies have sometimes been successfully administered, which combine with the carbonic acid gas, and thus at once reduce its volume. A flexible probang, or in default of it, a rattan, or grape-vine, with a knot on the end, may be gently forced down the gullet, and the gas thus permitted to escape.

HYDATID ON THE BRAIN.

The symptoms of this disease, known as turnsick, sturdy, staggers, water in the head, etc., are a dull, moping appearance, the sheep separating from the flock, a wandering and blue appearance of the eye, and sometimes partial or total blindness; the sheep appears unsteady in its walk, will sometimes stop suddenly and fall down, at others gallop across the field, and, after the disease has existed for some time, will almost constantly move round in a circle—there seems, indeed, to be an aberration of the intellect of the animal. These symptoms, though rarely all present in the same subject, are yet sufficiently marked to prevent any mistake as to the nature of the disease.

On examining the brain of sheep thus affected, what appears to be a watery bladder, called a hydatid, is found, which may be either small or of the size of a hen's egg. This hydatid, one of the class of entozoöns, has been termed by naturalists the *hydatis polycephalus cerebralis*, or many-headed hydatid of the brain; these heads being irregularly distributed on the surface of the bladder, and on the front part of each head there is a mouth surrounded by minute sharp hooks within a ring of sucking disks. These disks serve as the means of attachment, by forming a vacuum, and bring the mouth in contact with the surface, and thus, by the aid of the hooks, the parasite is nourished. The coats of the hydatid are disposed in several layers, one of which appears to possess a muscular power. These facts are developed by the microscope, which also discloses numerous little bodies adhering to the internal membrane. The fluid in the bladdes is usually clear but occasionally turbid, and then it has been found to contain a number of minute worms.

Treatment. This is deemed an almost incurable disorder. Where the hydatid is not imbedded in the brain, its constant pressure, singularly enough, causes a portion of the cranium to be absorbed, and finally the part immediately over the hydatid becomes thin and soft enough to yield under the pressure of the finger.

When such a spot is discovered, the English veterinarians usually dissect back the muscular integuments, remove a portion of the bone, carefully divide the investing membranes of the brain, and then, if possible, remove the hydatid whole; or, failing to do this, remove its fluid contents. The membranes and integuments are then restored to their position, and an adhesive plaster placed over the whole. The French veterinarians usually simply puncture the cranium and the cyst with a trochar, and laying the sheep on its back, allow the fluid to run out through the orifice thus made. A common awl would answer every purpose for such a puncture; and the puncture is the preferable method for the unskilled practitioner. An instance is, indeed, recorded of a cure having been effected, where the animal had been given up, by boring with a gimlet into the soft place on the head, when the water rushed out, and the sheep immediately followed the others to the pasture.

When, however, the hazard and cruelty attending the operation, under the most favorable circumstances, are considered, as well as the conceded liability of a return of the malady—the growth of new hydatids—it is evident that in this country, it would not be worth while, except in the case of uncommonly valuable sheep, to adopt any other remedy than depriving the miserable animal of life.

OBSTRUCTION OF THE GULLET.

After pouring a little oil in the throat, the obstructing substance which occasions the "choking," can frequently be removed up or down by external manipulation. If not, it may usually be forced down with the flexible probang, described in "Cattle and their Diseases," or a flexible rod, the head of which is guarded by a knot, or a little bag of flax-seed. The latter having been dipped in hot water for a minute or two, is partly converted into mucilage, which constantly exudes through the cloth, and protects the œsophagus, or gullet, from laceration. But little force must be used, and the whole operation conducted with the utmost care and gentleness; or the œsophagus will be so far lacerated as to produce death, although the obstruction is removed.

A BARRACK FOR STORING SHEEP-FODDER.

OPHTHALMIA.

Ophthalmia, or inflammation of the eyes, is not uncommon in this country; but it is little noticed, as, in most cases, it disappears in a few days, or, at worst, is only followed by cataract, which, being usually confined to one eye, does not appreciably effect the value of the animal, and therefore has no influence on its market price.

Treatment. Some recommend blowing pulverized red chalk in the inflamed eye; others squirt into it tobacco juice. As a matter of humanity, blood may be drawn from under the eye, and the eye bathed in tepid water, and occasionally with a weak solution of the sulphate of zinc combined with tincture of opium. These latter applications diminish the pain, and hasten the cure.

PALSY.

Paralysis, or palsy, is a diminution or entire loss of the powers of motion in some parts of the body. In the winter, poor lambs, or poor pregnant ewes, or poor feeble ewes immediately after yeaning in the spring, occasionally lose the power of walking or standing rather too suddenly to have it referable to increasing debility. The animal seems to have lost all strength in its loins, and the hind-quarters are powerless; it makes ineffectual attempts to rise, and cannot stand if placed upon its feet.

Treatment. Warmth, gentle stimulants, and good nursing may raise the patient; but, in the vast majority of cases, it is more economical and equally humane, to deprive it of life at once.

PELT-ROT.

This is often mistaken for the scab, but it is, in fact, a different and less dangerous disease. The wool falls off, and leaves the sheep nearly naked; but it is attended with no soreness, though a redish crust will cover the skin, from the wool which has dropped. It generally arises from hard keeping and much exposure to cold and wet; and, in fact, the animal often dies in severe weather from the cold it suffers on account of the loss of its coat.

The *remedy* is full feeding, a warm stall, and anointing the hard part of the skin with tar, oil, and butter. Some, however, do nothing for it, scarcely considering it a disease. Such say that if the condition of a poor sheep is raised as suddenly as practicable, by generous keep in the winter, the wool is very apt to drop off; and, if yet cold, the sheep will require warm shelter.

PNEUMONIA.

Pneumonia—or inflammation of the lungs—is not a common disease in the Northern States; but undoubted cases of it sometimes occur, after sheep have been exposed to sudden cold, particularly when recently shorn. The adhesions occasionally witnessed between the lungs and pleura of slaughtered sheep, betray the former existence of this disease in the animal—though, in many instances, it was so slight as to be mistaken, at the time, for a hard cold.

Symptoms. The animal is dull, ceases to ruminate, neglects its food, drinks frequently and largely, and its breathing is rapid and laborious; the eye is clouded; the nose discharges a tenacious, fetid matter; th~ teeth are ground frequently, so that the sound is audible at some distance; the pulse is at first hard and rapid, sometimes intermits, but before death it becomes weak. During the height of the fever, the flanks heave violently; there is a hard, painful cough during the first stages, which becomes weaker, and seems to be accompanied with more pain as death approaches.

After death, the lungs are found more or less hepatized—that is, permanently condensed and engorged with blood, so that their structure resembles that of the *hepar*, or liver—and they have so far lost their integrity that they are torn asunder by the slightest force. It may here be remarked that when sheep die from any cause, *with their blood in them*, the lungs have a dark, hepatized appearance. Whether they are actually hepatized or not, can readily be decided by compressing the wind-pipe, so that air cannot escape through it, and then between such compression and the body of the lungs, in a closely fitting orifice, inserting a goose-quill, or other tube, and continuing to blow until the lungs are inflated as far as they can be. As they inflate, they will become of a lighter color, and plainly manifest their cellular structure. If any portions of them cannot be inflated, and retain their dark, liver-like consistence, and color, they exhibit hepatization—the result of high inflammatory action—and a state utterly incompatible, in the living animal, with the discharge of the natural functions of the viscus.

Treatment. In the first, or inflammatory stages, bleeding and aperients are clearly called for. Some recommend early and copious bleeding, repeated, if necessary, in a few hours; this followed by aperient medicines, such as two ounces of

Epsom salts, which may be repeated in smaller doses, if the bowels are not sufficiently relaxed. The following sedative may also be given with gruel, twice a day: nitrate of potash, one drachm; powdered digitalis, one scruple; and tartarized antimony, one scruple.

While depletion may be of inestimable value during the continuance—the short continuance—of the febrile state, yet excitation like this will soon be followed by corresponding exhaustion, when the bleeding and purging would be murderous expedients; and gentian, ginger, and the spirits of nitrous ether will afford the only hope of cure.

POISON.

Sheep will often, in the winter or spring, eat greedily of the low laurel. The animal appears afterward to be dull and stupid, swells a little, and is constantly gulping up a feverish fluid, which it swallows again; a part of it will trickle out of its mouth, and discolor its lips. The plant probably brings on a fermentation in the stomach, and nature endeavors to throw off the poisonous herb by retching or vomiting.

Treatment. In the early stages, if the greenish fluid be allowed to escape from the stomach, the animal generally recovers. To effect this, gag the sheep, which may be done in this manner: Take a stick of the size of the wrist, six inches long—place it in the animal's mouth—tie a string to one end of it, pass it over the head and down to the other end, and there make it fast. The fluid will then run from the mouth as fast as thrown up from the stomach. In addition to this, give roasted onions and sweetened milk freely. A better plan, however, is to force a gill of melted lard down the throat; or, boil for an hour the twigs of the white ash, and give one-half to one gill of the strong liquor immediately; to be repeated, if not successful. Drenchers of milk and castor-oil are also recommended.

ROT.

This disease, which sometimes causes the death of a million of sheep, in England, in a single year, is comparatively unknown in this country. It prevails somewhat in the Western States, from allowing sheep to pasture on land that is overflowed with water. Even a crop of green oats, early in the fall, before a frost comes, has been known to rot young sheep.

Symptoms. The first are by no means strongly marked; there is no loss of condition, but rather the contrary, to all appearance. A paleness and want of liveliness of the membranes, generally, may be considered as the first symptoms, to which may be added a yellowness of the caruncle at the corner of the eye. When in warm, sultry, or rainy weather, sheep that are grazing on low and moist lands, feed rapidly, and some of them die suddenly, there is ground for fearing that they have contracted the rot. This suspicion will be farther increased if, a few days afterward, the sheep begin to shrink and grow flaccid about the loins. By pressure about the hips at this time, a crackling is perceptible now or soon afterward, the countenance looks pale, and upon parting the fleece, the skin is found to have changed its vermilion tint for a pale red, and the wool is easily separated from the felt; and as the disorder advances, the skin becomes dappled with yellow or black spots. To these symptoms succeed increased dullness, loss of condition, and greater paleness of the mucous membranes, the eyelids becoming almost white, and afterward yellow. This yellowness extends to other parts of the body, and a watery fluid appears under the skin, the latter becoming loose and flabby, and the wool coming off readily. The symptoms of dropsy often extend over the body, and sometimes the sheep becomes *chockered*, as it is termed; a large swelling forms under the jaw, which, from the appearance of the fluid which it contains, is sometimes called the *watery poke*. The duration of the disease is uncertain; the animal occasionally dies shortly after becoming affected, but more frequently it extends to from three to six months, the sheep gradually losing flesh and pining away, particularly if, as is frequently the case, an obstinate purging supervenes.

Post-mortem. The whole cellular tissue is found to be infiltrated, and a yellow serous fluid everywhere follows the knife. The muscles are soft and flabby, having the appearance of being macerated. The kidneys are pale, flaccid, and infiltrated. The mesenteric glands are enlarged, and engorged with yellow serous fluid. The belly is frequently filled with water, or purulent matter; the peritoneum is everywhere thickened, and the bowels adhere together by means of an unnatural growth The heart is enlarged and softened, and the lungs are filled with tubercles. The principal alterations of structure are in the liver, which is pale, livid, and broken down with the slightest pressure; and on being boiled, it will almost dissolve away. When the liver is not pale, it is often curiously spotted; in some cases it is speckled, like the back of a toad; some parts of it, however, are hard and schirrous; others are ulcerated, and the biliary ducts are filled with flukes. The malady is, unquestionably, inflammation of the liver.

This fluke is from three-quarters of an inch to an inch and a quarter in length, and from one-third to one-half an inch in its greatest breadth. These fluke-worms undoubtedly aggravate the disease, and perpetuate a state of irritability and disorganization, which must necessarily undermine the strength of any animal.

Treatment. This must, to a considerable extent, be very unsatisfactory. After the use of dry food and dry bedding, one of the best *preventives* is the abundant use of pure salt. In violent attacks, take eight, ten, or twelve ounces of blood, according to the circumstances of the case ; to this, let a dose of physic succeed—two or three ounces of Epsom salts ; and to these means add a change of diet, good hay in the field, and hay, straw, or chaff in the yard. After the operation of the physic—an additional dose having been administered, oftentimes, in order to quicken the action of the first—two or three grains of calomel may be given daily, mixed with half the quantity of opium, in order to secure its beneficial, and ward off its injurious effects on the ruminant. To this should be added common salt, which acts as a purgative and a tonic A mild tonic, as well as an aperient, is plainly indicated soon. after the commencement of rot. The doses should be from two to three drachms, repeated morning and night. When the inflammatory stage is clearly passed, stronger tonics may be added to the salt, and there are none superior to the gentian and ginger roots ; from one to two drachms of each, finely pounded, may be added to each dose of the salt. The sheep having a little recovered from the disease, should still continue on the best and driest pasture on the farm, and should always have salt within their reach. The rot is not infectious.

SCAB.

This is a cutaneous disease, analogous to the mange in horses and the itch in man, and is caused and propagated by a minute insect, the *acarus*.

If one or more female *acari* are placed on the wool of a sound sheep, they quickly travel to the root of it, and bury themselves in the skin, the place at which they penetrate being scarcely visible, or only distinguishable by a minute red point. On the tenth or twelfth day, a little swelling may be detected with the finger, and the skin changes its color, and has a greenish blue tint. The pustule is now rapidly formed, and about the sixteenth day breaks, when the mothers again appear, with their little ones attached to their feet, and covered by a portion of the shell of the egg from which they have just escaped. These little ones immediately set to work, penetrate the neighboring skin, bury themselves beneath it, find their proper nourishment, and grow and propagate, until the poor creature has myriads of them preying upon him. It is not wonderful that, under such circumstances, he should speedily sink. The male *acari*, when placed on the sound skin of a sheep, will likewise burrow their way and

THE BROAD-TAILED SHEEP.

disappear for a while, the pustule rising in due time; but the itching and the scab soon disappear without the employment of any remedy. The female brings forth from eight to fifteen young at a time.

In the United States, this disease is comparatively little known, and never originates spontaneously. The fact, that short-woolled sheep—like the Merino—are much less subject to its attacks, is probably one reason for this slight comparative prevalence. The disease spreads from individual to individual, and from flock to flock, not only by means of direct contact, but by the *acari* left on posts, stones, and other substances against which diseased sheep have rubbed themselves. Healthy sheep are, therefore, liable to contract the malady, if turned on pastures previously occupied by scabby sheep. although some considerable time may have elapsed since the departure of the latter.

The sheep laboring under the scab is exceedingly restless. It rubs itself with violence against trees, stones, fences, etc.; scratches itself with its feet, bites its sores, and tears off its wool with its teeth; as the pustules are broken, their matter escapes, and forms scabs, causing red, inflamed sores, which constantly extend, increasing the misery of the tortured animal; if unrelieved, he pines away, and soon perishes.

The *post-mortem* appearances are very uncertain and inconclusive. There is generally chronic inflammation of the intestines, with the presence of a great number of worms. The liver is occasionally schirrous, and the spleen enlarged; and there are frequently serous effusions in the belly, and sometimes in the chest. There has been evident sympathy between the digestive and the cutaneous systems.

Treatment. First, separate the sheep; then cut off the wool as far as the skin feels hard to the finger; the scab is then washed with soap-suds, and rubbed hard with a shoe-brush, so that it may be cleansed and broken. For this use take a decoction of tobacco, to which add one-third, by measure, of the lye of wood-ashes, as much hog's lard as will be dissolved by the lye, a small quantity of tar from a tar-bucket, which contains grease, and about one-eighth of the whole, by measure, of spirits of turpentine. This liquor is rubbed upon the part infected, and spread to a little distance around it, in three washings, with an interval of three days each. This will invariably effect a cure, when the disorder is only partial.

Or, the following: Dip the sheep in an infusion of arsenic, in the proportion of half a pound of arsenic to twelve gallons of water. The sheep should be previously washed in soap and water. The infusion must not be permitted to enter the mouth or nostrils.

Or, take common mercurial ointment; for bad cases, rub it down with three times its weight of lard—for ordinary cases, five times its weight. Rub a little of this ointment into the head of the sheep. Part the wool so as to expose the skin in a line from the head to the tail, and then apply a little of the ointment with the finger the whole way. Make a similar furrow and application on each side, four inches from the first; and so on, over the whole body. The quantity of ointment after composition with the lard, should not exceed two ounces; and, generally, less will suffice. A lamb requires but one-third as much as a grown sheep. This will generally cure; but, if the animal should continue to rub itself, a lighter application of the same should be made in ten days.

Or, take two pounds of lard or palm oil; half a pound of oil of tar; and one pound of sulphur; gradually mix the last two, then rub down the compound with the first. Apply as before. Or, take of corrosive sublimate, one half a pound; white hellabore, powdered, three-fourths of a pound; whale or other oil, six gallons; rosin, two pounds; and tallow, two pounds. The first two to be mixed with a little of the oil, and the rest being melted together, the whole to be gradually mixed. This is a powerful preparation, and must not be applied too freely.

An erysipelatous scab, or erysipelas, attended with considerable itching, sometimes troubles sheep. This is a febrile disease, and is treated with a cooling purgative, bleeding, and oil or lard applied to the sores.

SMALL-POX.

The author acknowledges himself indebted for what follows under this head to R. McClure, V.S., of Philadelphia, author of a Prize Essay on Diseases of Sheep, read before the U. S. Agricultural Society, in 1860, for which a medal and diploma were awarded.

Although the small-pox in domestic animals has, fortunately, been as yet confined to the European Continent—where it has been chiefly limited to England—no good reason can ever be assigned why it should not at some future time make its appearance among us, especially when we remember how long a period elapsed, during which we escaped the cattle plague, although the Continent had long been suffering from it.

The small-pox in sheep—*variola overia*—is, at times, epizoötic in the flocks of France and Italy, but was unknown in England until 1847, when it was communicated to a flock at Datchett and another at Pinnier by some Merinos from Spain. It soon found its way into Hampshire and Norfolk, but was shortly afterward supposed to be eradicated. In 1862, however, it suddenly reappeared in a severe form among the flocks of Wiltshire; for which reappearance neither any traceable infection nor contagion could be assigned. With the present light upon the subject, it would seem to be an instance of the origination *anew* of a malignant type of varioloid disease. Such an origin is, in fact, assigned to this disease in Africa, it being well established that certain devitalizing atmospheric influences produce skin diseases, and facilitate the appearance of pustular eruptions.

The disease once rooted soon becomes epizoötic, and causes a greater mortality than any other malady affecting this animal. Out of a flock numbering 1720, 920 were attacked in a natural way, of which 50 per cent. died. Of 800 inoculated cases, but 36 per cent. died.

Numerous experiments have proved beyond all doubt that this disease in sheep is both infectious and contagious; its period of incubation varies from seven to fourteen days. The mortality is never less than 25 per cent., and not unfrequently whole flocks have been swept away, death taking place in the early stages of the eruption, or in the stages of suppuration and ulceration.

The *symptoms* may be mapped out as follows: The animal is seized with a shivering fit, succeeded by a dull stupidity, which remains until death or recovery results; on the second or third day, pimples are seen on the thighs and arm-pits, accompanied with extreme redness of the eyes, complete loss of appetite, etc., etc. It is needless to enumerate other symptoms which exist in common with those of other disorders.

Prevention. At present, but two modes are resorted to, for the purpose of preventing the spread of the disease, which promise any degree of certainty of success. The first is by *inoculation*, which was recommended by Professor Simonds, of London. This distinguished pathologist appears to have overlooked the fact that he was thereby only enlarging the sphere of mischief, by imparting the disease to animals that, in all probability, would otherwise have escaped it. By inoculation, moreover, a form of the disease is given, not of a modified character, but with all the virulence of the original affection which is to be arrested, and equally as potent for further destruction of others. By such teaching, inoculation and vaccination would be made one and the same thing, notwithstanding their dissimilarity. Even vaccination will not protect the animal, as has been already shown by the experiments of Hurbrel D'Arboval.

The second and best plan of prevention is *isolation and destruction*, as recommended by Professor Gamgee, of the Edinburgh Veterinary College. This proved a great protection to the sheep-farmers of Wiltshire, in 1862. In all epizoötic diseases, individual cases occur, which, when pointed out and recognized as soon as the fever sets in and the early eruptions appear, should be slaughtered at once and buried, and the rest of the flock isolated. By this means the disease has been confined to but two or three in a large flock.

Treatment. In treating this disease, resort has of late been had to a plant, known as *Sarracenia purpura*—Indian cup, or pitcher plant—used for this purpose by the Micmacs, a tribe of Indians in British North America. This plant is indigenous, perennial, and is found from the coast of Labrador to the shores of the Gulf of Mexico, growing in great abundance on wet, marshy ground. The use of this plant is becoming quite general, and good results have almost uniformly attended it.

Take from one to two ounces of the dried root, and slice in thin pieces; place in an earthen pot; add a quart of cold water, and allow the liquid to simmer gently over a steady fire for two or three hours, so as to lose one-fourth of the quantity. Give of this decoction three wine-glassfuls at once, and the same quantity from four to six hours afterwards, when a cure will generally be affected. Weaker and smaller doses are certain preventives of the disease. The public are indebted to Dr. Morris, physician to the Halifax (Nova Scotia) Dispensary, for the manner of preparing this eminently useful article.

SORE FACE.

Sheep feeding on pastures infested with John's wort, frequently exhibit an irritation of skin about the nose and face, which causes the hair to drop off from the parts. The irritation sometimes extends over the entire body. If this plant is eaten in too large quantities, it produces violent inflammation of the bowels, and is frequently fatal to lambs, and sometimes to adults.

Treatment. Rub a little sulphur and lard on the irritated surface. If there are symptoms of inflammation of the bowels, this should be put into the mouth of the sheep with a flattened stick. Abundance of salt is deemed a *preventive.*

SORE MOUTH.

The lips of sheep sometimes become suddenly sore in the winter, and swell to the thickness of a man's hand. The malady occasionally attacks whole flocks, and becomes quite fatal. It is usually attributed to noxious weeds cut with the hay.

Treatment. Daub the lips and mouth plentifully with tar.

TICKS.

The treatment necessary as a preventive against these insects, and a remedy for them, has already been indicated under the head of "FEEDING AND MANAGEMENT," to which the reader is referred.

THE hog is a cosmopolite, adapting itself to almost every climate; though its natural haunts—like those of the hippopotamus, the elephant, the rhinoceros, and most of the thick-skinned animals—are in warm countries. They are most abundant in China, the East Indies, and the immense range of islands extending throughout the whole Southern and Pacific oceans; but they are also numerous throughout

Europe, from its Southern coast to the Russian dominions within the Arctic.

As far back as the records of history extend, this animal appears to have been known, and his flesh made use of as food. Nearly fifteen hundred years before Christ, Moses gave those laws to the Israelites which have given rise to so much discussion; and it is evident that, had not pork been the prevailing food of that nation at the time, such stringent commandments and prohibitions would not have been necessary. The various allusions to this kind of meat, which repeatedly occur in the writings of the old Greek authors, show the esteem in which it was held among that nation; and it appears that the Romans made the art of breeding, rearing, and fattening pigs a study. In fact, the hog was very highly prized among the early nations of Europe; and some of the ancients even paid it divine honors.

The Jews, the Egyptians, and the Mohammedans alone appear to have abstained from the flesh of swine. The former were expressly denied its use by the laws of Moses. "And the swine, though he divide the hoof, and be cloven-footed, yet he cheweth not the cud; he is unclean unto you." Lev. xi. 7. Upon this prohibition, Mohammed, probably, founded his own. For the Mosaic prohibition, various reasons have been assigned: the alleged extreme filthiness of the animal; it being afflicted with a leprosy; the great indigestibility of its flesh in hot climates; the intent to make the Jews "a peculiar people;" a preventive of gluttony; and an admonition of abstinence from sensual and disgusting habits.

At what period the animal was reclaimed from his wild state, and by what nation, cannot be stated. From the earliest times, in England, the hog has been regarded as a very important animal, and vast herds were tended by swineherds, who watched over their safety in the woods, and collected them under shelter at night. Its flesh was the staple article of consumption in every household, and much of the wealth of the rich and free portion of the community consisted in these animals. Hence bequests of swine, with land for their support, were often made; rights and privileges connected with their feeding, and the extent of woodland to be occupied by a given number, were granted according to established rules. Long after the end of the Saxon dynasty, the practice of feeding swine upon the mast and acorns of the forest was continued till the forests were cut down, and the land laid open for the plough.

Nature designed the hog to fulfil many important functions in a forest country. By his burrowing after roots and the like, he turns up and destroys the larvæ of innumerable insects, which would otherwise injure the trees as well as their fruit. He destroys the slug-snail and adder, and thus not only rids the forests of these injurious and unpleasant inhabitants, but also makes them subservient to his own nourishment, and therefore to the benefit of mankind. The fruits which he eats are such as would otherwise rot on the ground and be wasted, or yield nutriment to vermin; and his diggings for earth-nuts and the like, loosens the soil, and benefits the roots of the trees. Hogs in forest land may, therefore, be regarded as eminently beneficial; and it is only the abuse which is to be feared.

The hog is popularly regarded as a stupid, brutal, rapacious, and filthy animal, grovelling and disgusting in all his habits, intractable and obstinate in temper. The most offensive epithets among men are borrowed from him, or his peculiarities. In their native state, however, swine seem by no means destitute of natural affections; they are gregarious, assemble together in defence of each other, herd together for warmth, and appear to have feelings in common; no mother is more tender to her young than the sow, or more resolute in their defence. Neglected as this animal has ever been by authors, recorded instances are not wanting of their sagacity, tractability, and susceptibility of affection. Among the European peasantry, where the hog is, so to speak, one of the family, he may often be seen following his master from place to place, and grunting his recognition of his protectors.

The hog, in point of actual fact, is also a much more cleanly animal than he has the credit of being. He is fond of a good, cleanly bed; and when this is not provided for him, it is oftentimes interesting to note the degree of sagacity with which he will forage for himself. It is, however, so much the vogue to believe that he may be kept in any state of neglect, that the terms "pig," and "pig-sty" are usually regarded as synonymous with all that is dirty and disgusting. His rolling in the mud is cited as a proof of his filthy habits. This practice, which he shares in common with all the pachydermatous animals, is undoubtedly the teaching of instinct, and for the purpose of cooling himself and keeping off flies.

Pigs are exceedingly fond of comfort and warmth, and will nestle together in order to obtain the latter, and often struggle vehemently to secure the warmest berth. They are likewise peculiarly sensitive of approaching changes in the weather, and may often be observed suddenly leaving the places in which they had been quietly feeding, and running off to their styes at full speed, making loud outcries. When storms are overhanging, they collect straw in their mouths, and run about as if inviting their companions to do the same; and if there is a shed or shelter near at hand, they will carry it there and deposit it, as if for the purpose of preparing a bed.

In their domesticated state, they are, undeniably, very greedy animals; eating is the business of their lives; nor do they appear to be very delicate as to the kind or quality of food which is placed before them. Although naturally herbivorous animals, they have been known to devour carrion with all the voracity of beasts of prey, to eat and mangle infants, and even gorge their appetites with their own young. It is not, however, unreasonable to believe that the last revolting act—rarely if ever happening in a state of nature—arises more from the pain and irritation produced by the state of confinement, and often filth, in which the animal is kept, and the disturbances to which it is subjected, than from any actual ferocity; for it is well known that a sow is always unusually irritable at this period, snapping at all animals that approach her. If she is gently treated, properly supplied with sustenance, and sequestered from all annoyance, there is little danger of this practice ever happening.

All the offences which swine commit are attributed to a disposition innately bad; whereas they too often arise from bad management, or total neglect. They are legitimate objects for the sport of idle boys, hunted with curs, pelted with stones, often neglected and obliged to find a meal for themselves, or wander about half-starved. Made thus the Ishmaelites of our domestic animals, is it a matter of wonder that they should, under such circumstances, incline to display Ishmaelitish traits? In any well-regulated farm-yard, the swine are as tractable and as little disposed to wander or trespass as any of the animals that it contains.

The WILD BOAR is generally admitted to be the parent of the stock from which all our domesticated breeds and varieties have sprung. This animal is generally of a dusky brown or iron-gray color, inclining to black, and diversified with black spots or streaks: The body is covered with coarse hairs, intermixed with a downy wool; these hairs become bristles as they approach the neck and shoulders, and are in those places so long as to form a mane, which the animal erects when irritated. The head is short, the forehead broad and flat, the ears short, rounded at the tips, and inclined toward the neck, the jaw armed with sharp, crooked tusks, which curve slightly upward, and are capable of inflicting fearful wounds, the eye full, neck thick and muscular, the shoulders high, the loins broad, the tail stiff, and finished off with a tuft of bristles at the tip, the haunch well turned, and the leg strong. A full-grown wild boar in India averages from thirty to forty inches in height at the shoulder; the African wild boar is about twenty-eight or thirty inches high.

The wild boar is a very active and powerful animal, and becomes fiercer as he grows older. When existing in a state of nature, he is generally found in moist, shady, and well-wooded situations, not far remote from streams or water. In India, they are found in the thick jungles, in plantations of sugar-cane or rice, or in the thick patches of high, long grass. In England, France, Germany, Italy, and Spain, their resorts have been in the woods and forests. This animal is naturally herbivorous, and appears to feed by choice upon plants, fruits, and roots. He will, however, eat the worms and larvæ which he finds in the ground, also snakes and other such reptiles, and the eggs of birds. They seldom quit their coverts during the day, but prowl about in search of food during twilight and the night. Their acute sense of smell enables them to detect the presence of roots or fruits deeply imbedded in the soil, and they often do considerable mischief by ploughing up the ground in search of them, particularly as they do not, like the common hog, root up a little spot here and there, but plough long, continuous, furrows.

The wild boar, properly so called, is neither a solitary nor a gregarious animal. For the first two or three years, the whole herd follows the sow, and all unite in defence against any enemies, calling upon each other with loud cries in case of emergency, and forming in regular line of battle, the weakest occupying the rear. When arrived at maturity, the animals wander alone, as if in perfect consciousness of their strength, and appear as if they neither sought nor avoided any living creature. They are reputed to live about thirty years; as they grow old, the hair becomes gray, and the tusks begin to show symptoms of decay. Old boars rarely associate with a herd, but seem to keep apart from the rest, and from each other.

The female produces but one litter in the year, much smaller in number than those of the domestic pig; she carries her young sixteen or twenty weeks, and generally is only seen with the male during the rutting season. She suckles her young for several months, and continues to protect them for some time afterward; if attacked at that time, she will defend herself and them with exceeding courage and fierceness. Many sows will often be found herding together, each followed by her litter of young; and in such parties they are exceedingly formidable to man and beast. Neither they nor the boar, however, seem desirous of attacking any thing; and only when roused by aggression, or disturbed in their retreat, do they turn upon their enemies and manifest the mighty strength with which Nature has endowed them. When attacked by dogs, the wild boar at first sullenly retreats, turning upon them from time to time and menacing them with his tusks; but gradually his anger rises, and at length he stands at bay, fights furiously for his life, and tears and rends his persecutors. He has even been observed to single out the most tormenting of them, and rush savagely upon him. Hunting this animal has been a favorite sport, in almost all countries in which it has been found, from the earliest ages.

THE WILD BOAR AT BAY.

Wild boars lingered in the forests of England and Scotland for several centuries after the Norman conquest, and many tracts of land in those countries derived their name from this circumstance; while instances of valor in their destruction are recorded in the heraldic devices of many of their noble families. The precise period at which the animal became exterminated there cannot be precisely ascertained. They had, however, evidently been long extinct in the time of Charles I., since he endeavored to re-introduce them, and was at considerable expense to procure a wild boar and his mate from Germany. They still exist in Upper Austria, on the Syrian Alps, in many parts of Hungary, and in the forests of Poland, Spain, Russia, and Sweden; and the inhabitants of those countries hunt them with hounds, or attack them with fire-arms, or with the proper boar-spear.

All the varieties of the domestic hog will breed with the wild boar; the period of gestation is the same in the wild and the tame sow; their anatomical structure is identical; their general form bears the same characters; and their habits, so far as they are not changed by domestication, remain the same. Where individuals of the pure wild race have been caught young and subjected to the same treatment as a domestic pig, their fierceness has disappeared, they have become more social and less nocturnal in their habits, lost their activity, and lived more to eat. In the course of one or two generations, even the form undergoes certain modifications; the body becomes larger and heavier; the legs shorter, and less adapted for exercise; the formidable tusks of the boar, being no longer needed as weapons of defence, disappear; the shape of the head and neck alters; and in character as well as in form, the animal adapts itself to its situation. Nor does it appear that a return to their native wilds restores to them their original appearance; for, in whatever country pigs have escaped from the control of man, and bred in the wilderness and woods, not a single instance is on record in which they have resumed the habits and form of the wild boar. They, indeed, become fierce, wild, gaunt, and grisly, and live upon roots and fruits; but they are, notwithstanding, merely degenerated swine, and they still associate together in herds, and do not walk solitary and alone, like their grim ancestors.

AMERICAN SWINE.

In the United States, swine have been an object of attention since its earliest settlement, and whenever a profitable market has been found for pork abroad, it has been exported to the full extent of the demand. Swine are not, however, indigenous to this country, but were doubtless originally brought hither by the early English settlers; and the breed thus introduced may still be distinguished by the traces they retain of their parent stock. France, also, as well as Spain, and, during the existence of the slave-trade, Africa, have also combined to furnish varieties of this animal, so much esteemed throughout the whole of the country, as furnishing a valuable article of food. For nearly twenty years following the commencement of the general European wars, soon after the organization of our national government, pork was a comparatively large article of commerce; but exports for a time diminished, and it was not until within a more recent period that this staple has been brought up to its former standard as an article of exportation to that country. The recent use which has been made of its carcass in converting it into lard oil, has tended to still further increase its consumption. By the census of 1860, there were upward of thirty-two and a half millions of these animals in the United States.

They are reared in every part of the Union, and, when

properly managed, always at a fair profit. At the extreme North, in the neighborhood of large markets, and on such of the Southern plantations as are particularly suited to sugar or rice, they should not be raised beyond the number required for the consumption of the coarse or refuse food produced. Swine are advantageously kept in connection with a dairy or orchard; since, with little additional food besides what is thus afforded, they can be put in good condition for the butcher.

On the rich bottoms and other lands of the West, however, where Indian corn is raised in profusion and at small expense, they can be reared in the greatest numbers and yield the largest profit. The Scioto, Miami, Wabash, Illinois, and other valleys, and extensive tracts in Kentucky, Tennessee, Missouri, and some adjoining States, have for many years taken the lead in the production of Swine; and it is probable that the climate and soil, which are peculiarly suited to their rapid growth, as well as that of their appropriate food, will enable them to hold their position as the leading pork-producers of the North American Continent.

The breeds cultivated in this country are numerous; and, like our native cattle, they embrace many of the best, and a few of the worst, to be found among the species. Great attention has been paid, for many years, to their improvement in the Eastern States; and nowhere are there better specimens than in many of their yards. This spirit has rapidly extended West and South; and among most of the intelligent farmers, who make them a leading object of attention, on their rich corn-grounds, swine have attained a high degree of excellence. This does not consist in the introduction and perpetuity of any distinct races, so much as in the breeding up to a desirable size and aptitude for fattening, from such meritorious individuals of any breed, or their crosses, as come within their reach.

THE BYEFIELD.

This breed was formerly in high repute in the Eastern States, and did much good among the species generally. They are white, with fine curly hair, well made and compact, moderate in size and length, with broad backs, and at fifteen months attaining some three hundred to three hundred and fifty pounds net.

THE BEDFORD.

The Bedford or Woburn is a breed originating with the Duke of Bedford, on his estate at Woburn, and brought to their perfection, probably, by judicious crosses of the Chinese hog on some of the best English swine. A pair was sent by the duke to this country, as a present to General Washington; but they were dishonestly sold by the messenger, in Maryland, in which State, and in Pennsylvania, they were productive of much good at an early day, by their extensive distribution through different States. Several other importations of this breed have been made at various times, and especially by the enterprising masters of the Liverpool packets, in the neighborhood of New York. They are a large, spotted animal, well made, and inclining to early maturity and fattening. This is an exceedingly valuable hog, but nearly extinct, both in England and in this country, as a breed.

THE LEICESTER.

The old Leicestershire breed, in England, was a perfect type of the original hogs of the midland counties; large, ungainly, slab-sided animals, of a light color, and spotted with brown or black. The only good parts about them were their heads and ears, which showed greater traces of breeding than any other portions. These have been materially improved by various crosses, and the original breed has nearly lost all its peculiarities and defects. They may now be characterized as a large, white hog, generally coarse in the bone and hair, great eaters, and slow in maturing. Some varieties differ essentially in these particulars, and mature early on a moderate amount of food. The crosses with small compact breeds are generally thrifty, desirable animals.

THE YORKSHIRE.

The old Yorkshire breed was one of the very large varieties, and one of the most unprofitable for a farmer, being greedy feeders, difficult to fatten, and unsound in constitution. They were of a dirty white or yellow color, spotted with black, had long legs, flat sides, narrow backs, weak loins, and large bones. Their hair was short and wirey, and intermingled with numerous bristles about the head and neck, and their ears long. When full grown and fat, they seldom weighed more than from three hundred and fifty to four hundred pounds.

These have been crossed with pigs of the improved Leicester breed; and where the crossings have been judiciously managed, and not carried too far, a fine race of deep-sided. short-legged, thin-haired animals has been obtained, fattening kindly, and rising to a weight of from two hundred and fifty to four hundred pounds, when killed between one and two years old; and when kept over two years, reaching even from five hundred to seven hundred pounds.

They have also been crossed with the Chinese, Neapolitan, and Berkshire breeds, and hardy, profitable, well-proportioned animals thereby obtained. The original breed, in its purity, size, and defectiveness, is now hardly to be met with, having shared the fate of the other large old breeds, and given place to smaller and more symmetrical animals. The *Yorkshire white* is among the large breeds deserving commendation among us. To the same class belong also the large *Miami white*, and the *Kenilworth;* each frequently attaining, when dressed, a weight of from six hundred to eight hundred pounds.

THE CHINESE.

This hog is to be found in the southeastern countries of Asia, as Siam, Cochin China, the Burman Empire, Cambodia, Malacca, Sumatra, and in Batavia, and other Eastern islands; and is, without doubt, the parent stock of the best European and American swine.

There are two distinct varieties, the *white* and the *black;* both fatten readily, but from their diminutive size attain no great weight. They are small in limb, round in body, short in the head, wide in the cheek, and high in the chime; covered with very fine bristles growing from an exceedingly thin skin; and not peculiarly symmetrical, since, when fat, the head is so buried in the neck that little more than the tip of the snout is visible. The pure Chinese is too delicate and susceptible to cold ever to become a really profitable animal in this country; it is difficult to rear, and the sows are not good nurses; but one or two judicious crosses have, in a manner, naturalized it. This breed will fatten readily, and on a comparatively small quantity of food; the flesh is exceedingly delicate, but does not make good bacon, and is often too fat and oily to be generally esteemed as pork. They are chiefly kept by those who rear sucking-pigs for the market, as they make excellent roasters at three weeks or a month old. Five, and even seven, varieties of this breed are distinguished, but these are doubtless the results of different crosses with our native kinds; among these are black, white, black and white, spotted, blue and white, and sandy.

THE CHINESE HOG.

Many valuable crosses have been made with these animals; for the prevalent fault of the old English breeds having been coarseness of flesh, unwieldiness of form, and want of aptitude to fatten, an admixture of the Chinese breed has materially corrected these defects. Most of our smaller breeds are more or less indebted to the Asiatic swine for their present compactness of form, the readiness with which they fatten on a small quantity of food, and their early maturity; but these advantages are not considered, in the judgment of some, as sufficiently great to compensate for the diminution in size, the increased delicacy of the animals, and the decrease of number in the litters. The best cross is between the Berkshire and Chinese.

THE SUFFOLK.

The old Suffolks are white in color, long-legged, long-bodied, with narrow backs, broad foreheads, short hams, and an abundance of bristles. They are by no means profitable animals. A cross between the Suffolk and Lincoln has produced a hardy animal, which fattens kindly, and attains the weight of from four hundred to five hundred and fifty, and even seven hundred pounds. Another cross, much approved by farmers, is that of the Suffolk and Berkshire.

THE SUFFOLK PIG.

There are few better breeds, perhaps, than the improved Suffolk—that is, the Suffolk crossed with the Chinese. The greater part of the pigs on the late Prince Albert's farm, near Windsor, were of this breed. They are well-formed, compact, of medium size, with round, bulky bodies, short legs, small heads, and fat cheeks. Many, at a year or fifteen months old, weigh from two hundred and fifty to three hundred pounds; at which age they make fine bacon hogs. The sucking-pigs are also very delicate and delicious.

Those arising from Berkshire and Suffolk are not so well shaped as the latter, being coarser, longer-legged, and more prominent about the hips. They are mostly white, with thin, fine hair; some few are spotted, and are easily kept in fine condition; they have a decided aptitude to fatten early, and are likewise valuable as store-pigs.

THE BERKSHIRE.

The Berkshire pigs belong to the large class, and are distinguished by their color, which is a sandy or whitish brown, spotted regularly with dark brown or black spots, and by their having no bristles. The hair is long, thin, somewhat curly, and looks rough; the ears are fringed with long hair round the outer edge, which gives them a ragged or feathery appearance; the body is thick, compact, and well formed; the legs short, the sides broad, the head well set on, the snout short, the jowl thick, the ears erect,

A BERKSHIRE BOAR.

skin exceedingly thin in texture, the flesh firm and well flavored, and the bacon very superior. This breed has generally been considered one of the best in England, on account of its smallness of bone, early maturity, aptitude to fatten on little food, hardihood, and the females being good breeders. Hogs of the pure original breed have been known to weigh from eight hundred to nine hundred and fifty pounds.

Numerous crosses have been made from this breed; the principal foreign ones are those with the Chinese and Neapolitan swine, made with the view of decreasing the size of the animal, improving the flavor of the flesh, and rendering it more delicate; and the animals thus attained are superior to almost any others in their aptitude to fatten; but are very susceptible to cold, from being almost entirely without hair. A cross with the Suffolk and Norfolk also is much improved, which produces a hardy kind, yielding well when sent to the butcher; although, under most circumstances, the pure Berkshire is the best.

No other breeds have been so extensively diffused in the United States, within comparatively so brief a period, as the Berkshires, and they have produced a marked improvement in many of our former races. They weigh variously, from two hundred and fifty to four hundred pounds net, at sixteen months, according to their food and style of breeding; and some full-grown have dressed to more than eight hundred pounds. They particularly excel in their hams, which are round, full, and heavy, and contain a large proportion of lean, tender, and juicy meat, of the best flavor.

None of our improved breeds afford long, coarse hair or bristles; and it is a gratifying evidence of our decided improvement in this department of domestic animals, that our brush-makers are obliged to import most of what they use from Russia and northern Europe. This improvement is manifest not only in the hair, but in the skin, which is soft and mellow to the touch; in the finer bones, shorter head, upright ears, dishing face, delicate muzzle, and wild eye; and in the short legs, low flanks, deep and wide chest, broad back, and early maturity.

THE NATURAL HISTORY OF THE HOG.

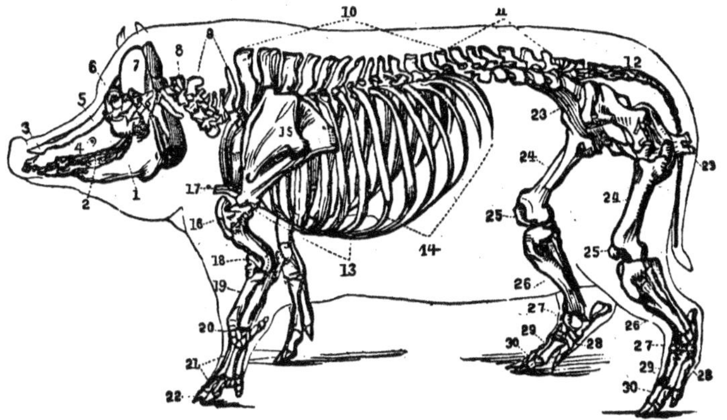

SKELETON OF THE HOG AS COVERED BY THE MUSCLES.

1. The lower jaw. 2. The teeth. 3. The nasal bones. 4. The upper jaw. 5. The frontal bone. 6. The orbit or socket of the eye. 7. The occipital bone. 8. The first vertebræ of the neck. 9. The vertebræ of the neck 10. The vertebræ of the back. 11. The vertebræ of the loins. 12. The bones of the tail. 13, 14 The true and false ribs. 15. The shoulder-blade. 16. The round shoulder-bone 17. The breast-bone. 18. The elbow. 19. The bone of the fore-arm. 20. The navicular bone. 21. The first and second bones of the foot. 22. The bones of the hoof. 23. The haunch bones. 24. The thigh bone. 25. The stifle bone. 26. The upper bone of the leg. 27. The hock bones. 28. The navicular bone. 29. The first digits of the foot. 30. The second digits of the foot.

DIVISION. *Vertebrata*—possessing a back-bone.

CLASS. *Mammalia*—such as give suck.

ORDER *Pachydermata*—thick-skinned.

FAMILY. *Suidæ*—the swine kind.

GENUS. *Sus*—the hog. Of this genus there are five varieties.

Sus Scropa, or Domestic Hog.

Sus Papuensis, or Bene.

Sus Guineensis, or Guinea Hog.

Sus Africanus, or Masked Boar.

Sus Babirussa, or Babirussa.

A very slight comparison of the face of this animal with that of any other will prove that strength is the object in view—strength toward the inferior part of the bone. In point of fact, the snout of the hog is his spade, with which, in his natural state, he digs and ruts in the ground for roots, earth-nuts, worms, etc. To render this implement more nearly perfect, an extra bone is added to the nasal bone, being connected with it by strong ligaments, cartilages, and muscles, and termed the snout-bone, or spade-bone, or ploughshare. By it and its cartilaginous attachment, the snout is rendered strong as well as flexible, and far more efficient than it otherwise could be; and the hog often continues to give both farmers and gardeners very unpleasant proofs of its efficiency, by ploughing up deep furrows in newly-sown fields, and grubbing up the soil in all directions in quest of living and dead food.

As roots and fruits buried in the earth form the natural food of the hog, his face terminates in this strong, muscular snout, insensible at the extremity, and perfectly adapted for turning up the soil. There is a large plexus or fold of nerves proceeding down each side of the nose; and in these, doubtless, resides that peculiar power which enables the hog to select his food, though buried some inches below the surface of the ground. The olfactory nerve is likewise large, and occupies a middle rank between that of the herbivorous and carnivorous animals; it is comparatively larger than that of the ox; indeed, few animals—with the exception of the dog, none—are gifted with a more acute sense of smell than the hog. To it epicures are indebted for the truffles which form such a delicious sauce, for they are the actual finders. A pig is turned into a field, allowed to pursue his own course, and watched. He stops, and begins to grub up the earth; the man hurries up, drives him away, and secures the truffle, which is invariably growing under that spot; and the poor pig goes off to sniff out another, and another, only now and then being permitted, by way of encouragement, to reap the fruits of his research.

FORMATION OF THE TEETH.

The hog has fourteen *molar* teeth in each jaw, six *incisors*, and two *canines;* these latter are curved upward, and commonly denominated *tushes.* The molar teeth are all slightly different in structure, and increase in size from first to last; they bear no slight resemblance to those of the human being. The incisors are so fantastic in form that they cannot well be described, and their destined functions are by no means clear. Those in the lower jaw are long, round, and nearly straight; of those in the upper jaw, four closely resemble the corresponding teeth in the horse; while the two corner incisors bear something of the shape of those of the dog. These latter are placed so near the tushes as often to obstruct their growth, and it is sometimes necessary to draw them, in order to relieve the animal and enable him to feed.

The hog is born with two molars on each side of the jaw; by the time he is three or four months old, he is provided with his incisive milk-teeth and the tushes; the supernumerary molars protrude between the fifth and seventh month, as does the first back molar; the second back molar is cut at about the age of ten months; and the third, generally, not until the animal is three years old. The upper corner teeth are shed at about the age of six or eight months; and the lower ones at about seven, nine, or ten months old, and replaced by the permanent ones. The milk tushes are also shed and replaced between six and ten months old. The age of twenty months, and from that to two years, is denoted by the shedding and replacement of the middle incisors, or *pincers*, in both jaws, and the formation of a black circle at the base of each of the tushes. At about two years and a half or three years of age, the adult middle teeth in both jaws protrude, and the pincers are becoming black and rounded at the ends.

After three years, the age may be computed by the growth of the tushes; at about four years, or rather before, the upper tushes begin to raise the lip; at five, they protrude through the lips; and at six years, the tushes of the lower jaw begin to show themselves out of the mouth, and assume a spiral form. These acquire a prodigious length in old animals, and particularly in uncastrated boars; and as they increase in size, they become curved backward and outward, and at length are so crooked as to interfere with the motion of the jaws to such a degree that it is necessary to cut off those projecting teeth, which is done with the file, or with nippers.

In the selection of a boar and sow for breeding, much more attention and consideration are requisite than is generally imagined. It is as easy, with a very little judgment and management, to procure a good as an inferior breed; and the former is much more remunerative, in proportion to the outlay, than the latter can possibly ever be.

The object of the farmer or breeder is to produce and retain such an animal as will be best adapted to the purpose he has in view, whether that is the consumption of certain things which could not otherwise be so well disposed of, the converting into hams, bacon, and pork, or the raising of sucking-pigs and porkers for the market. Almost all farmers keep one or more pigs to devour the offal and refuse, which would otherwise be wasted. This is, however, a matter totally distinct from breeding swine. In the former case, the animal or animals are purchased young for a small price, each person buying as many as he considers he shall have food enough for, and then sold to the butcher, or killed, when in proper condition; and thus a certain degree of profit is realized. In the latter, many contingencies must be taken into account: the available means of feeding them; whether or not the food may be more profitably disposed of; the facilities afforded by railroads, the vicinity of towns, or large markets, etc., for disposing of them.

In the breeding of swine, as much as that of any other live-stock, it is important to pay great attention, not only to the breed, but also to the choice of individuals. The sow should produce a great number of young ones, and she must be well fed to enable her to support them. Some sows bring forth ten, twelve, or even fifteen pigs at a birth; but eight or nine is the usual number; and sows which produce fewer than this must be rejected. It is, however, probable that fecundity depends also on the boar; he should, therefore, be chosen from a race which multiplies quickly.

If a bacon and a late market be objects, the large and heavy varieties should be selected, care being taken that the breed has the character of possessing those qualities most likely to insure a heavy return—growth, and facility of taking fat. Good one-year bacon-hogs being in great demand, they may be known by their long bodies, low bellies, and short legs. With these qualities are usually coupled long, pendulous ears, which attract purchasers. If, however, hogs are to be sold at all seasons to the butchers, the animals must attain their full growth and be ready for killing before they are a year old. This quality is particularly prominent in the Chinese breed; but among our ordinary varieties, hogs are often met with better adapted for this purpose than for producing large quantities of bacon and lard. The Berkshire crossed with Chinese is an excellent porker.

The sow should be chosen from a breed of proper size and shape, sound and free from blemishes and defects. In every case—whether the object be pork or bacon—the *points* to be looked for in the *sow* are a small, lively head; a broad and deep chest; round ribs; capacious barrel; a haunch falling almost to the hough; deep and broad loin; ample hips; and considerable length of body, in proportion to its height. One qualification should ever be kept in view, and, perhaps, should be the first point to which the attention should be directed—that is, smallness of bone. She should have at least twelve teats; for it is observed that each pig selects a teat for himself and keeps to it, so that a pig not having one belonging to him would be starved. A good sow should produce a great number of pigs, all of equal vigor. She must be very careful of them, and not crush them by her weight; above all, she must not be addicted to eating the after-birth, and, what may often follow, her own young. If a sow is tainted with those bad habits, or if she has difficult labors, or brings forth dead pigs, she must be spayed forthwith. It is, therefore, well to bring up several young sows at once, so as to keep those only which are free from defects. Breeding sows and boars should never be raised from defective animals. Sows that have very low bellies, almost touching the ground, seldom produce large or fine litters. A good-sized sow is generally considered more likely to prove a good breeder and nurse, and to farrow more easily and safely than a small, delicate animal.

The ancients considered the distinguishing marks of a good *boar* to be a small head, short legs, a long body, large thighs and neck, and this latter part thickly covered with strong, erect bristles. The most experienced modern breeders prefer an animal with a long, cylindrical body; small bones; well-developed muscles; a wide chest, which denotes strength of constitution; a broad, straight back; short head and fine snout; brilliant eyes; a short, thick neck; broad, well-developed shoulders; a loose, mellow skin; fine, bright, long hair, and few bristles; and small legs and hips. Some give the preference to long, flapping ears; but experience seems to demonstrate that those animals are best which have short, erect, fine ears. The boar should always be vigorous and masculine in appearance.

Few domesticated animals suffer so much from in-and-in breeding as swine. Where this system is pursued, the number of young ones is decreased at every litter, until the sows become, in a manner, barren. This practice also undoubtedly contributes to their liability to hereditary diseases, such as scrofula, epilepsy, and rheumatism; and when those possessing any such diseases are coupled, the ruin of the flock is easily and speedily effected, since they are propagated by either parent, and always most certainly and in most aggravated form, when occurring in both. As soon as the slightest degeneracy is observed, the breed should be crossed from time to time, keeping sight, however, while so doing, of the end in view. The Chinese will generally be found the best which can be used for this purpose; since a single cross, and even two, with one of these animals, will seldom do harm, but often effect considerable improvements. The best formed of the progeny resulting from this cross must be selected as breeders, and with them the old original stock crossed back again. Selection, with judicious and cautious admixture, is the true secret of forming and improving the breed. Repeated and indiscriminate crosses are as injurious as an obstinate adherence to one particular breed, and as much to be avoided.

The following rules for the selection of the best stock of hogs will apply to all breeds:

Fertility. In a breeding sow, this quality is essential, and it is one which is inherited. Besides this, she should be a careful mother. A young, untried sow will generally display in her tendencies those which have predominated in the race from which she has descended. Both boar and sow should be sound, healthy, and in fair, but not over fat, condition.

Form Where a farmer has an excellent breed, but with certain defects, or too long in the limb, or too heavy in the bone, the sire to be chosen, whether of a pure or of a cross breed, should exhibit the opposite qualities, even to an extreme; and be, moreover, one of a strain noted for early and rapid fattening. If in perfect health, young stock selected for breeding will be lively, animated, hold up the head, and move freely and nimbly.

Bristles. These should be fine and scanty, so as to show the skin smooth and glossy; coarse, wirey, rough bristles usually accompany heavy bones, large, spreading hoofs, and flapping ears, and thus become one of the indications of a thick-skinned and low breed.

Color. Different breeds of high excellence have their own colors; white, black, parti-colored, black and white, sandy, mottled with large marks of black, are the most prevalent. A black skin, with short, scanty bristles, and small stature, demonstrate the prevalence of the Neapolitan strain, or the black Chinese, or, perhaps, an admixture of both. Many prefer white; and in sucking-pigs, destined for the table, and for porkers, this color has its advantages, and the skin looks more attractive; it is, however, generally thought that the skin of black hogs is thinner than that of white, and less subject to eruptive diseases.

The influence of a first impregnation upon subsequent progeny by other males is at times curiously illustrated. This has been noticed in respect of the sow. A sow of the black and white breed, in one instance, became pregnant by a boar of the wild breed of a deep chestnut color. The pigs produced were duly mixed, the color of the boar being very predominant in some. The sow being afterwards put to a boar of the same breed as herself, some of the produce were still stained or marked with the chestnut color which prevailed in the first litter; and the same occurred after a third impregnation, the boar being then of the same kind as herself. What adds to the force of this case is, that in the course of many years' observation, the breed in question was never known to produce progeny having the slightest tinge of chestnut color.

A sow is capable of conceiving at the age of six or seven months; but it is always better not to let her commence breeding too early, as it tends to weaken her. From ten to twelve months—and the latter is preferable—is about the best age. The boar should be, at least, a twelvemonth old—some even recommend eighteen months, at least—before he is employed for the purpose of propagating his species. If, however, the sow has attained her second year, and the boar his third, a vigorous and numerous offspring is more likely to result. The boar and sow retain their ability to breed for almost five years; that is, until the former is upward of eight years old, and the latter seven. It is not advisable, however, to use a boar after he has passed his fifth year, nor a sow after her fourth, unless she has proved a peculiarly valuable breeder—in which case she might produce two or three more litters.

A boar left on the pasture, at liberty with the sows, might suffice for thirty or forty of them; but as he is commonly shut up, and allowed access at stated times only, so that the young ones may be born at nearly the same time, it is usual to allow him to serve from six to ten—on no account should he serve more. The best plan is, to shut up the boar and sow in a sty together; for, when turned in among several females, he is apt to ride them so often that he exhausts himself without effect. The breeding boar should be fed well and kept in high condition, but not fat. Full grown boars being often savage and difficult to tame, and prone to attack men and animals, should be deprived of their tusks.

Whenever it is practicable, it should always be so arranged

that the animals shall farrow early in the spring, and at the latter end of summer, or quite the beginning of autumn. In the former case, the young pigs will have the run of the early pastures, which will be a benefit to them, and a saving to their owners; and there will also be more whey, milk, and other dairy produce which can be spared for them by the time they are ready to be weaned. In the second case, there will be sufficient time for the young to have grown and acquired strength before the cold weather comes on, which is always very injurious to sucking-pigs.

POINTS OF A GOOD HOG.

It may be not amiss to group together what is deemed desirable under this head. No one should be led away by mere name in his selection of a hog. It may be called a Berkshire, or a Suffolk, or any other breed most in estimation, and yet, in reality, may possess none of this valuable blood. The only sure way to avoid imposition is, to make *name* always secondary to *points.* If a hog is found possessing such points of form as are calculated to insure early maturity and faculty of taking on flesh, one needs to care but little by what name he is called; since no mere name can bestow value upon an animal deficient in the qualities already indicated.

The true Berkshire—that possessing a dash of the Chinese and Neapolitan varieties—comes, perhaps, nearer to the desired standard than any other.

The chief points which characterize such a hog are the following:—In the first place, sufficient depth of carcass, and such an elongation of body as will insure a sufficient lateral expansion. The loin and breast should be broad. The breadth of the former denotes good room for the play of the lungs, and, as a consequence, a free and healthy circulation, essential to the thriving or fattening of any animal. The bone should be small, and the joints fine—nothing is more indicative of high breeding than this; and the legs should be no longer than, when fully fat, would just prevent the animal's belly from trailing upon the ground. The leg is the least profitable portion of the hog, and no more of it is required than is absolutely necessary for the support of the rest. The feet should be firm and sound; the toes should lie well together, and press straightly upon the ground; the claws, also, should be even, upright and healthy.

The form of the head is sometimes deemed of little or no consequence, it being generally, perhaps, supposed that a good hog may have an ugly head; but the head of all animals is one of the very principal points in which pure or impure breeding will be most obviously indicated. A high-bred animal will invariably be found to arrive more speedily at maturity, to take flesh more easily, and at an earlier period, and, altogether, to turn out more profitably than one of questionable or impure stock. Such being the case, the head of the hog is a point by no means to be overlooked. The description of head most likely to promise—or, rather to be the accompaniment of—high breeding, is one not carrying heavy bones, not too flat on the forehead, or possessing a snout too elongated; the snout should be short, and the forehead rather convex, curving upward; and the ear, while pendulous, should incline somewhat forward, and at the same time be light and thin. The carriage of the pig should also be noticed. If this be dull, heavy, and dejected, one may reasonably suspect ill health, if not some concealed disorder actually existing, or just about to break forth; and there cannot be a more unfavorable symptom than a hung-down, slouching head. Of course, a fat hog for slaughter and a sow heavy with young, have not much sprightliness of deportment.

Color is, likewise, not to be disregarded. Those colors are preferable which are characteristic of the most esteemed breeds. If the hair is scant, black is desirable, as denoting connection with the Neapolitan; if too bare of hair, a too intimate alliance with that variety may be apprehended, and a consequent want of hardihood, which—however unimportant, if pork be the object—renders such animals a hazardous speculation for store purposes, on account of their extreme susceptibility of cold, and consequent liability to disease. If white, and not too small, they are valuable as exhibiting connection with the Chinese. If light, or sandy, or red with black marks, the favorite Berkshire is detected; and so on, with reference to every possible variety of hue.

TREATMENT DURING PREGNANCY.

Sows with pigs should be well and judiciously fed; that is to say, they should have a sufficiency of wholesome, nutritious food to maintain their strength and keep them in good condition, but should by no means be allowed to get fat; as when they are in high condition, the dangers of parturition are enhanced, the animal is more awkward and liable to smother and crush her young, and, moreover, never has as much or as good milk as a leaner sow. She should also have a separate sty; for swine are prone to lie so close together that, if she is even among others, her young would be in great danger; and this sty should be perfectly clean and comfortably littered, but not so thickly as to admit of the young being able to bury themselves in the straw.

As the time of her farrowing approaches, she should be well supplied with food, especially if she be a young sow, and this her first litter, and also carefully watched, in order to prevent her devouring the after-birth, and thus engendering a morbid appetite which will next induce her to fall upon her own young. A sow that has once done this can never afterward be depended upon. Hunger, thirst, or irritation of any kind, will often induce this unnatural conduct, which is another reason why a sow about to farrow should have a sty to herself, and be carefully attended to, and have all her wants supplied.

ABORTION.

This is by no means of so common occurrence in the case of the sow as in many other of the domesticated animals. Various causes tend to produce it: insufficiency of food, eating too much succulent vegetable food, or unwholesome, unsubstantial diet; blows and falls; and the animal's habit of rubbing itself against hard bodies, for the purpose of allaying the irritation produced by the vermin or cutaneous eruptions to which it is subject. Reiterated copulation does not appear to produce abortion in the sow; at least to the extent it does in other animals.

The symptoms indicative of approaching abortion are similar to those of parturition, but more intense. There are, generally, restlessness, irritation, and shivering; and the cries of the animal evince the presence of severe labor-pains.

Sometimes the rectum, vagina, or uterus, becomes relaxed, and one or the other protrudes, and often becomes inverted at the moment of the expulsion of the fœtus, preceded by the placenta, which presents itself foremost.

Nothing can be done, at the last hour, to prevent abortion; but, from the first, every predisposing cause should be removed. The treatment will depend upon circumstances. Where the animal is young, vigorous, and in high condition, bleeding will be beneficial—not a copious blood-letting, but small quantities taken at different times; purgatives may also be administered. If, when abortion has taken place, the whole of the litter was not born, emollient injections may be resorted to with considerable benefit; otherwise, the after treatment should be made the same as in parturition, and the animal should be kept warm, quiet, and clean, and allowed a certain degree of liberty. Whenever one sow has aborted, the causes likely to have produced this accident should be sought, and an endeavor made, by removing them, to secure the rest of the inmates of the piggery from a similar mishap.

In cases of abortion, the fœtus is seldom born alive, and often has been dead for some days; where this is the case—which may be readily detected by a peculiarly unpleasant putrid exhalation, and the discharge of a fetid liquid from the vagina—the parts should be washed with a diluted solution of chloride of lime, in the proportion of one part of chloride to three parts of water, and a portion of this lotion gently injected into the uterus, if the animal will submit to it. Mild doses of Epsom salts, tincture of gentian, and Jamaica ginger, will also act beneficially in such cases, and, with attention to diet, soon restores the animal.

PARTURITION.

The period of gestation varies according to age, constitution, food, and the peculiarities of the individual breed. The most usual period during which the sow carries her young is, according to some, three months, three weeks, and three days,

WILD HOGS.

or one hundred and eight days; according to others, four lunar months, or sixteen weeks, or about one hundred and thirteen days. It may safely be said to range from one hundred and nine to one hundred and forty-three days.

The sow produces from eight to thirteen young at a litter, and sometimes even more. Young and weakly sows not only produce fewer pigs, but farrow earlier than those of maturer age and sounder condition; and besides, as might be expected, their offspring are deficient in vigor, oftentimes, indeed, puny and feeble. Extraordinary fecundity is not however, desirable, for nourishment cannot be afforded to more than twelve, the sow's number of teats. The supernumerary pigs must therefore suffer; if but one, it is, of course, the smallest and weakest; a too numerous litter are all, indeed, generally undersized and weakly, and seldom or never prove profitable; a litter not exceeding ten will usually be found to turn out most advantageously. On account of the discrepancy between the number farrowed by different sows, it is a good plan, if it can be managed, to have more than one breeding at the same time, in order that the number to be suckled by each may be equalized. The sow seldom recognizes the presence of a strange little one, if it has been introduced among the others during her absence, and has lain for half an hour or so among her own offspring in their sty.

The approach of the period of farrowing is marked by the immense size of the belly, by a depression of the back, and by the distention of the teats. The animal manifests symptoms of acute suffering, and wanders restlessly about, collecting straw, and carrying it to her sty, grunting piteously meanwhile. As soon as this is observed, she should be persuaded into a separate sty, and carefully watched. On no account should several sows be permitted to farrow in the same place at the same time, as they will inevitably irritate each other, or devour their own or one another's young.

The young ones should be taken away as soon as they are born, and deposited in a warm spot; for the sow being a clumsy animal, is not unlikely in her struggles to overlie them; nor should they be returned to her, until all is over, and the after-birth has been removed, which should always be done the moment it passes from her; for young sows, especially, will invariably devour it, if permitted, and then, as the young are wet with a similar fluid, and smell the same, they will eat them also, one after another. Some advise washing the backs of young pigs with a decoction of aloes, colocynth, or some other nauseous substance, as a remedy for this; but the simplest and easiest one is to remove the little ones until all is over, and the mother begins to recover herself and seek about for them, when they should be put near her. Some also recommend strapping up the sow's mouth for the first three or four days, only releasing it to admit of her taking her meals.

Some sows are apt to lie upon and crush their young. This may best be avoided by not keeping her too fat or heavy, and by not leaving too many young upon her. The straw forming the bed should likewise be short, and not in too great quantity, lest the pigs get huddled up under it, and the sow unconsciously over-lie them in that condition.

It does not always happen that the parturition is effected with ease. Cases of false presentation, of enlarged fœtus, and of debility in the mother, often render it difficult and dangerous. The womb will occasionally become protruded and inverted, in consequence of the forcing pains of difficult parturition, and even the bladder has been known to come away. These parts must be returned as soon as may be; and if the womb has come in contact with the dung or litter, and acquired any dirt, it must first be washed in lukewarm water, and then returned, and confined in its place by means of a suture passed through the lips of the orifice. The easiest and perhaps the best way, however, is not to return the protruded parts at all, but merely tie a ligature round them and leave them to slough off, which they will do in the course of a few days, without effusion of blood, or farther injury to the animal. No sow that has once suffered from protrusion of the womb should be allowed to breed again.

TREATMENT WHILE SUCKLING.

Much depends upon this; as many a fine sow and promising litter have been ruined for want of proper and judicious care at this period. Immediately after farrowing, many sows incline to be feverish; where this is the case, a light and sparing diet only should be given them for the first day or two, as gruel, oatmeal porridge, whey, and the like. Others, again, are very much debilitated, and require strengthening; for them, strong soup, bread steeped in wine, or in a mixture of brandy and sweet spirits of nitre, administered in small quantities, will often prove highly beneficial.

The rations must gradually be increased and given more frequently; and they must be composed of wholesome, nutritious, and succulent substances. All kinds of roots—carrots, turnips, potatoes, and beet-roots—well steamed or boiled, but never raw, may be given; bran, barley, and oatmeal, bran-flour, Indian corn, whey, sour, skim, and butter-milk, are all well adapted for this period; and, should the animal appear to require it, grain well bruised and macerated may be added. Whenever it is possible, the sow should be turned out for an hour each day, to graze in a meadow or clover-field, as the fresh air, exercise, and herbage, will do her immense good. The young pigs must be shut up for the first ten days or fortnight, after which they will be able to follow her, and take their share of the benefit.

The food should be given regularly at certain hours; small and often-repeated meals are far preferable to large ones, since indigestion, or any disarrangement of the functions of the stomach vitiates the milk, and produces diarrhœa and other similar affections in the young. The mother should always be well fed, but not over-fed; the better and more carefully she is fed, the more abundant and nutritious will her milk be, the better will the sucking-pigs thrive, and the less will she be reduced by suckling them.

When a sow is weakly, and has not a sufficiency of milk, the young pigs must be taught to feed as early as possible. A kind of gruel, made of skim-milk and bran, or oatmeal, is a good thing for this purpose, or potatoes, boiled and then mashed in milk or whey, with or without the addition of a little bran or oatmeal. Toward the period when the pigs are to be weaned, the sow must be less plentifully fed, otherwise the secretion of milk will be as great as ever; it will, besides, accumulate, and there will be hardness, and perhaps inflammation of the teats. If necessary, a dose of physic may be given to assist in carrying off the milk; but, in general, a little judicious management in the feeding and weaning will be all that is required.

TREATMENT OF YOUNG PIGS.

For the first ten days, or a fortnight, the mother will generally be able to support her litter without assistance, unless, as has been already observed, she is weakly, or her young are too numerous; in either of which cases they must be fed from the first. When the young pigs are about a fortnight old, warm milk should be given to them. In another week, this may be thickened with some species of farina; and afterward, as they gain strength and increase in size, boiled roots and vegetables may be added. As soon as they begin to eat, an open frame or railing should be placed in the sty, under which the little pigs can run, and on the otherofe d so this should be the small troughs containing their food; for it never answers to let them eat out of the same trough with their mother, because the food set before her is generally too strong and stimulating for them, even if they should secure any of it, which is, to say the least, extremely doubtful. Those intended to be killed for sucking-pigs should not be above four weeks old; most kill them for this purpose on the twenty-first or twenty-second day. The others, excepting those kept for breeding, should be castrated at the same time.

CASTRATION AND SPAYING.

Pigs are chiefly castrated with a view to fattening them; and, doubtless, this operation has the desired effect—for at the same time that it increases the quiescent qualities of the animal, it diminishes also his courage, spirits, and nobler attributes, and even affects his form. The tusks of a castrated boar never grow like those of the natural animal, but always have a dwarfed, stunted appearance. The operation, if possible, should be performed in the spring or autumn, as the temperature is the more uniform, and care should be taken that the animal is in perfect health. Those which are fat and plethoric should be prepared by bleeding, cooling diet and quiet. Pigs are castrated at all ages, from a fortnight to three, six and eight weeks, and even four months old.

There are various modes of performing this operation. If the pig is not more than six weeks old, an incision is made at the bottom of the scrotum, the testicle pushed out, and the cord cut, without any precautionary means whatever. When the animal is older, there is reason to fear that hemorrhage, to a greater or less extent, will supervene; consequently, it will be advisable to pass a ligature round the cord a little above the spot where the division is to take place.

By another mode—to be practised only on very young animals—a portion of the base of the scrotum is cut off, the testicles forced out, and the cord sawn through with a somewhat serrated but blunt instrument. If there is any hemorrhage, it is arrested by putting ashes in the wound. The animal is then dismissed and nothing further done with him.

On animals two and three years old, the operation is sometimes performed in the following manner: An assistant holds the pig, pressing the back of the animal against his chest and belly, keeping the head elevated, and grasping all the four legs together; or, which is the preferable way, one assistant holds the animal against his chest, while another kneels down and secures the four legs. The operator then grasps the scrotum with his left hand, makes one horizontal incision across its base, opening both divisions of the bag at the same time. The testicles are then pressed out with his finger and thumb, and removed with a blunt knife, which lacerates the part without bruising it and rendering it painful. Laceration only is requisite in order to prevent the subsequent hemorrhage which would occur, if the cord were simply severed by a sharp instrument. The wound is then closed by pushing the edges gently together with the fingers, and it speedily heals. Some break the spermatic cord without tearing it; they twist it, and then pull it gently and finally until it gives way.

In other cases, a waxed cord is passed as tightly as possible round the scrotum, above the epididymus, which completely stops the circulation, and in a few days the scrotum and testicles will drop off. This operation should never be performed on pigs of more than six weeks of age, and the spermatic should always, first of all, be measured. It, moreover requires great nicety and skill; otherwise, accidents will occur, and considerable pain and inflammation be caused. Too thick a cord, a knot not tied sufficiently tight, or a portion of the testicle included in the ligature, will prevent its success.

The most fatal consequence of castration is tetanus, or lockjaw, induced by the shock communicated to the nervous system by the torture of the operation.

SPAYING.

This operation consists in removing the ovaries, and sometimes a portion of the uterus, more or less considerable, of the female. The animal is laid upon its left side, and firmly held by one or two assistants; an incision is then made into the flank, the forefinger of the right hand introduced into it, and gently moved about until it encounters and hooks hold of the right ovary, which it draws through the opening; a ligature is then passed round this one, and the left ovary felt for in like manner. The operator then severs these two ovaries, either by cutting or tearing, and returns the womb and its appurtenances to their proper position. This being done, he closes the wound with two or three stitches, sometimes rubs a little oil over it, and releases the animal. All goes on well. for the healing power of the pig is very great.

The after-treatment is very simple. The animals should be

well littered with clean straw, in styes weather-tight and thoroughly ventilated; their diet should be cared for; some milk or whey, with barley-meal is an excellent article; it is well to confine them for a few days, as they should be prevented from getting into cold water or mud until the wound is perfectly healed, and also from creeping through fences.

The best age for spaying a sow is about six weeks; indeed, as a general rule, the younger the animal is when either operation is performed the quicker it recovers. Some persons, however, have two or three litters from their sows before they operate upon them; where this is the case, the result is more to be feared, as the parts have become more susceptible, and are, consequently, more liable to take on inflammation.

WEANING.

Some farmers wean the pigs a few hours after birth, and turn the sow at once to the boar. The best mode, however, is to turn the boar into the hog-yard about a week after parturition, at which time the sow should be removed a few hours daily from her young. It does not injure either the sow or her pigs if she takes the boar while suckling; but some sows will not do so until the drying of their milk.

The age at which pigs may be weaned to the greatest advantage is when they are about eight or ten weeks old; many, however, wean them as early as six weeks, but they seldom turn out as well. They should not be taken from the sow at once, but gradually weaned. At first they should be removed from her for a certain number of hours each day, and accustomed to be impelled by hunger to eat from the trough; then they may be turned out for an hour without her, and afterwards shut up while she also is turned out by herself. Subsequently, they must only be allowed to suck a certain number of times in twenty-four hours; perhaps six times at first, then four, then three, and, at last, only once; and meanwhile they must be proportionably better and more plentifully fed, and the mother's diet in a like manner diminished. Some advise that the whole litter should be weaned at once; this is not best, unless one or two of the pigs are much weaker and smaller than the others; in such case, if the sow remain in tolerable condition, they might be suffered to suck for a week longer; but this should be the exception, and not a general rule.

Pigs are more easily weaned than almost any other animals, because they learn to feed sooner; but attention must, nevertheless, be paid to them, if they are to grow up strong, healthy animals. Their styes must be warm, dry, clean, well-ventilated, and weather-tight. They should have the run of a grass meadow or enclosure for an hour or two every fine day, in spring and summer, or be turned into the farm-yard among the cattle in the winter, as fresh air and exercise tend to prevent them from becoming rickety or crooked in the legs.

The most nutritious and succulent food that circumstances will permit should be furnished them. Newly-weaned pigs require five or six meals in the twenty-four hours. In about ten days, one may be omitted; in another week, a second; and then they should do with three *regular* meals each day. A little sulphur mingled with the food, or a small quantity of Epsom or Glauber salts dissolved in the water, will frequently prove beneficial. A plentiful supply of clear, cold water should always be within their reach; the food left in the trough after the animals have finished eating should be removed, and the trough thoroughly rinsed out before any more is put into it. Strict attention should also be paid to cleanliness. The boars and sows should be kept apart from the period of weaning.

The question, which is more profitable, to breed swine, or to buy young pigs and fatten them, can best be determined by those interested; since they know best what resources they can command, and what chance of profits each of these separate branches offers.

RINGING.

This operation is performed to counteract the propensity which swine have of digging and furrowing up the earth. The ring is passed through what appears to be a prolongation of the septum, between the supplemental, or snout-bone, and the nasal. The animal is thus unable to obtain sufficient purchase to use his snout with any effect, without causing the ring to press so painfully upon the part that he is forced to desist. The ring, however, is apt to break, or it wears out in process of time, and has to be replaced.

The snout should be perforated at weaning-time, after the animal has recovered from castration or spaying; and it will be necessary to renew the operation as it becomes of large growth. It is too generally neglected at first; but no pigs, young or old, should be suffered to run at large without this precaution. The sow's ring should be ascertained to be of sufficient strength previously to her taking the boar, on account of the risk of abortion, if the operation is renewed while she is with pig. Care must be taken by the operator not to go too close to the bone, and that the ring turn easily.

A far better mode of proceeding is, when the pig is young, to cut through the cartilaginous and ligamentous prolongations, by which the supplementary bone is united to the proper nasals. The divided edges of the cartilage will never re-unite, and the snout always remains powerless.

FEEDING AND FATTENING.

Roots and fruits are the natural food of the hog, in a wild as well as in a domesticated state; and it is evident that, however omnivorous it may occasionally appear, its palate is by no means insensible to the difference in eatables, since, whenever it finds variety, it will select the best with as much cleverness as other quadrupeds. Indeed, the hog is more nice in the selection of his vegetable diet than any of the other domesticated herbivorous animals. To a certain extent he is oiv-nm orous, and may be reared on the refuse of slaughter-houses; but such food is not wholesome, nor is it natural; for, though he is omnivorous, he is not essentially carnivorous. The refuse of the dairy-farm is more congenial to his health, to say nothing of the quality of its flesh.

Swine are generally fattened for pork at from six to nine months old; and for bacon, at from a year to two years. Eighteen months is generally considered the proper age for a good bacon hog. The feeding will always, in a great measure, depend upon the circumstances of the owner—upon the kind of food which he has at his disposal, and can best spare—and the purpose for which the animal is intended. It will also, in some degree, be regulated by the season; it being possible to feed pigs very differently in the summer from what they are fed in the winter.

The refuse wash and grains, and other residue of breweries and distilleries, may be given to swine with advantage, and seem to induce a tendency to lay on flesh. They should not, however, be given in too large quantities, nor unmixed with other and more substantial food; since, although they give flesh rapidly when fed on it, the meat is not firm, and never makes good bacon. Hogs eat acorns and beech-mast greedily, and so far thrive on this food that it is an easy matter to fatten them afterwards. Apples and pumpkins are likewise valuable for this purpose.

There is nothing so nutritious, so eminently and in every way adapted for the purpose of fattening, as are the various kinds of grain—nothing that tends more to create firmness as well as delicacy in the flesh. Indian corn is equal, if not superior, to any kind of grain for fattening purposes, and can be given in its natural state, as pigs are so fond of it that they will eat up every kernel. The pork and bacon of animals that have been thus fed are peculiarly firm and solid. Animal food tends to make swine savage and feverish, and often lays the foundation of serious inflammation of the intestines. Weekly washing with soap and a brush adds wonderfully to the thriving condition of a hog.

In the rich corn, regions of our States, upon that grain beginning to ripen, as it does in August, the fields are fenced off into suitable lots, and large herds are successively turned into them, to consume the grain at their leisure. They waste nothing except the stalks, which in that land of plenty are considered of little value, and they are still useful as manure for succeeding crops; and whatever grain is left by them, leaner droves which follow will readily glean. Peas, early buckwheat, and apples, may be fed on the ground in the same way.

There is an improvement in the character of the grain from a few months' keeping, which is fully equivalent to the interest of the money and the cost of storage. If fattened early in the season, hogs will consume less food to make an equal amount of flesh than in colder weather; they will require less attention; and, generally, early pork will command the highest price in market.

It is most economical to provide swine with a fine clover pasture, to run in during the spring and summer; and they ought also to have access to the orchard, to pick up all the unripe and superfluous fruit that falls. They should also have the wash of the house and the dairy, to which add meal, and let it sour in large tubs or barrels. Not less than one-third, and perhaps more, of the whole grain fed to hogs, is saved by grinding and cooking, or souring. Care must, however, be taken that the souring be not carried so far as to injure the food by putrefaction. A mixture of meal and water, with the addition of yeast or such remains of a former fermentation as adhere to the sides or bottom of the vessel, and exposure to a temperature between sixty-eight and seventy-seven degrees Fahrenheit, will produce immediate fermentation.

In this process there are five stages: the *saccharine*, by which the starch and gum of the vegetables, in their natural condition, are converted into sugar; the *vinous*, which changes the sugar into alcohol; the *mucilaginous*, sometimes taking the place of the vinous, and occurring where the sugar solution, or fermenting principle, is weak, producing a slimy, glutinous product; the *acetic*, forming vinegar, from the vinous or alcoholic stage; and the *putrefactive*, which destroys all the nutritive principles and converts them into a poison. The precise points in fermentation, when the food becomes most profitable for feeding, has not as yet been satisfactorily determined; but that it should stop short of the putrefactive, and probably the full maturity of the acetic, is certain.

The roots for fattening ought to be washed, and steamed or boiled; and when not intended to be fermented, the meal may be scalded with the roots. A small quantity of salt should be added. Potatoes are the best roots for swine; then parsnips; orange or red carrots, white or Belgian; sugar-beets; mangel-wurtzels; ruta-bagas; and then white turnips, in the order mentioned. The nutritive properties of turnips are diffused through so large a bulk that it is doubtful if they can ever be fed to fattening swine with advantage; and they will barely sustain life when fed to them uncooked.

There is a great loss in feeding roots to fattening swine, without cooking. When unprepared grain is fed, it should be on a full stomach, to prevent imperfect mastication, and consequent loss of the food. It is better, indeed, to have it always before them. The animal machine is an expensive one to keep in motion; and it should be the object of the farmer to put his food in the most available condition for its immediate conversion into fat and muscle.

The following injunctions should be rigidly observed, if one would secure the greatest results:

1. Avoid *foul feeding.*

2. Do not omit adding *salt* in moderate quantities to the mess given.

3. Feed at *regular intervals.*

4. *Cleanse* the troughs previous to feeding.

5. Do not *over-feed;* give only as much as will be consumed at the meal.

6. *Vary* the food. Variety will create, or, at all events, increase appetite, and it is most conducive to health. Let the variations be governed by the condition of the *dung* cast, which should be of a medium consistence, and of a grayish-brown color; if *hard,* increase the quantity of bran and succulent roots; if too *liquid,* diminish, or dispense with bran, and make the mess firmer; add a portion of corn.

7. Feed the stock *separately,* in classes, according to their relative conditions. Keep sows with young by themselves; store-hogs by themselves; and bacon-hogs and porkers by themselves. It is not advisable to keep the store-hogs too high in flesh, since high feeding is calculated to retard development of form and bulk. It is better to feed pigs intended to be put up for bacon *loosely* and not too abundantly, until they have attained their full stature; they can then be brought into the highest possible condition in a surprisingly short space of time.

8. Keep the swine *clean,* dry, and warm. Cleanliness, dryness, and warmth are *essential,* and as imperative as feeding; for an inferior description of food will, by their aid, succeed far better than the highest feeding will without them.

PIGGERIES.

Few items conduce more to the thriving and well-being of swine than airy, spacious, well-constructed styes, and above all, cleanliness. They were formerly too often housed in damp, dirty, close, and imperfectly-built sheds, which was a fruitful source of disease and of unthrifty animals. Any place was once thought good enough to keep a pig in.

In large establishments, where numerous pigs are kept, there should be divisions appropriated to all the different kinds; the boars, the breeding sows, the newly weaned, and the fattening pigs should all be kept separate; and in the divisions assigned to the second and last of these classes, it is best to have a distinct apartment for each animal, all opening into a yard or inclosure of limited extent. As pigs require warmth, these buildings should face the south, and be kept weather-tight and well drained. Good ventilation is also important; for it is idle to expect animals to make good flesh and retain their health, unless they have a sufficiency of pure air. The blood requires this to give it vitality and free it from impurities, as much as the stomach requires wholesome and strengthening food; and when it does not have it, it becomes vitiated, and impairs all the animal functions. Bad smells and exhalations, moreover, injure the flavor of the meat.

Damp and cold floors should be guarded against, as they tend to induce cramp and diarrhœa; and the roof should be so contrived as to carry off the wet from the pigs. The walls of a well-constructed sty should be of solid masonry; the roof sloping, and furnished with spouts to carry off the rain; the floors either slightly inclined toward a gutter made to carry off the rain, or else raised from the ground on beams or joists, and perforated so that all urine and moisture shall drain off. Bricks and tiles, sometimes used for flooring, are objectionable, because, however well covered with straw, they still strike cold. Wood is far superior in this respect, as well as because it admits of those clefts or perforations being made, which serve not only to drain off all moisture, but also to admit fresh air.

The manure proceeding from the pig-sty has often been much undervalued, and for this reason, that the litter is supposed to form the principal portion of it; whereas it constitutes the least valuable part, and, indeed, it can scarcely be regarded as manure at all—at least by itself—where the requisite attention is paid to the cleanliness of the animals and of their dwellings. The urine and the dung are valuable, being, from the very nature of the food of the animals, exceedingly rich and oleaginous, and materially beneficial to cold soils and grass-lands. The manure from the sty should always be collected as carefully as that from the stable or cow-house, and husbanded in the same way.

The door of each sty ought to be so hung that it will open inward or outward, so as to give the animals free ingress and egress. For this purpose, it should be hung across from side to side, and the animal can push it up to effect its entry or exit; for, if it were hung in the ordinary way, it would derange the litter every time it opened inward, and be very liable to hitch. If it is not intended that the pigs shall leave their sty, there should be an upper and lower door; the former of which should always be left open when the weather is warm and dry, while the latter will serve to confine the animal.

There should likewise be windows or slides, which can be opened or closed at will, to give admission to the fresh air, or exclude rain or cold.

Wherever it can be managed, the troughs—which should be of stone or cast metal, since wooden ones will soon be gnawed to pieces—should be so situated that they can be filled and cleaned from the outside, without interfering with or disturbing the animals at all; and for this purpose it is well to have a flap, or door, with swinging hinges, made to hang horizontally on the trough, so that it can be moved to and fro, and alternately be fastened by a bolt to the inside or outside of the manger. When the hogs have fed sufficiently, the door is swung inward and fastened, and so remains until feeding-time, when the trough is cleansed and refilled without any trouble, and then the flap drawn back, and the animals admitted to their food. Some cover the trough with a lid having as many holes in it as there are pigs to eat from it, which gives each pig an opportunity of selecting his own hole, and eating away without interfering with or incommoding his neighbor.

A hog ought to have three apartments, one each for sleeping, eating, and evacuations; of which the last may occupy the lowest, and the first the highest level, so that nothing shall be drained, and as little carried into the first two as possible. The piggery should always be built as near as possible to that portion of the establishment from which the chief part of the provision is to come, since much labor will thus be saved. Washings, and combings, and brushings, as has been previously suggested, are valuable adjuncts in the treatment of swine; the energies of the skin are thus roused, the pores opened, the healthful functions aided, and that inertness, so likely to be engendered by the lazy life of a fattening pig, counteracted.

A supply of fresh water is essential to the well-being of swine, and should be freely furnished. If a stream can be brought through the piggery, it answers better than any thing else. Swine are dirty feeders and dirty drinkers, usually plunging their forefeet into the trough or pail, and thus polluting with mud or dirt whatever may be given to them. One of the advantages, therefore, to be derived from the stream of running water is, its being kept constantly clean and wholesome by its running. If this advantage cannot be procured, it is desirable to present water in vessels of a size to receive but one head at a time, and of such height as to render it impossible, or difficult, for the drinker to get his feet into it. The water should be renewed twice daily. If swine are closely confined in pens, they should have as much charcoal twice a week as they will eat, for the purpose of correcting any tendency to disorders of the stomach. Rotten wood is an imperfect substitute for charcoal.

SLAUGHTERING.

A pig that is to be killed should be kept without food for from twelve to sixteen hours previous to slaughtering; a little water must, however, be within his reach. He should, in the first place, be stunned by a blow on the head. Some advise that the knife should be thrust into the neck so as to sever the artery leading from the heart; while others prefer that the animal should be stuck through the brisket in the direction of the heart—care being exercised not to toucht he first rib. The blood should then be allowed to drain from the carcass into vessels placed for the purpose; and the more completely it does so, the better will be the meat.

A large tub, or other vessel, has been previously got ready, which is now filled with boiling water. The carcass of the hog is plunged into this, and the hair is then removed with the edge of a knife. The hair is more easily removed if the hog is

THE OLD ENGLISH HOG.

scalded before he stiffens, or becomes quite cold. It is not, however, necessary, but simply brutal and barbarous, to scald him while there is yet some life in him. Bacon-hogs may be singed, by enveloping the body in straw, and setting the straw on fire, and then scraping it all over. When this is done, care must be observed not to burn or parch the cuticle. The entrails should then be removed, and the interior of the body well washed with lukewarm water, so as to remove all blood and impurities, and afterward wiped dry with a clean cloth; the carcass should then be hung up in a cool place for eighteen or twenty hours, to become set and firm.

For cutting up, the carcass should be laid on the back, upon a strong table. The head should then be cut off close by the ears, and the hinder feet so far below the houghs as not to disfigure the hams, and leave room sufficient for hanging them up; after which the carcass is divided into equal halves, up the middle of the back bone, with a cleaving-knife, and, if necessary, a hand-mallet. Then cut the ham from the side by the second joint of the back-bone, which will appear on dividing the carcass, and dress the ham by paring a little off the flank, or skinny part, so as to shape it with a half round point, clearing off any top fat which may appear. Next cut off the sharp edge along the back bone with a knife and mallet, and slice off the first rib next the shoulder, where there is a bloody vein, which must be taken out, since, if it is left in, that part is apt to spoil. The corners should be squared off when the ham is cut. The ordinary practice is to cut out the spine, or back bone. Some take out the chine and upper parts of the ribs in the first place; indeed, almost every locality has its peculiar mode of proceeding.

PICKLING AND CURING.

The usual method of curing is to pack the pork in clean salt, adding brine to the barrel when filled. But it may be dry-salted, by rubbing it in thoroughly on every side of each piece, with a strong leather rubber firmly secured to the palm of the right hand. The pieces are then thrown into heaps and sprinkled with salt, and occasionally turned till cured; or it may at once be packed in dry casks, which are rolled at times to bring the salt into contact with every part.

Hams and shoulders may be cured in the same manner either dry or in pickle, but with differently arranged materials.

The following is a good pickle for two hundred pounds: Take fourteen pounds of Turk's Island salt; one-half pound of saltpetre; two quarts of molasses, or four pounds of brown sugar; with water enough to dissolve them. Bring the liquor to the scalding-point, and skim off all the impurities which rise to the top. When cold, pour it upon the ham, which should be perfectly cool, but not frozen, and closely packed; if not sufficient to cover it, add pure water for this purpose. Some extensive packers of choice hams add pepper, allspice, cinnamon, nutmegs, or mace and cloves.

The hams may remain six or eight weeks in this pickle, then should be hung up in the smoke-house, with the small end down, and smoked from ten to twenty days, according to the quantity of smoke. The fire should not be near enough to heat the hams. In Holland and Westphalia, the fire is made in the cellar, and the smoke carried by a flue into a cool, dry chamber. This is, undoubtedly, the best mode of smoking. The hams should at all times be dry and cool, or their flavor will suffer. Green sugar-maple chips are best for smoke; next to them are hickory, sweet birch, corn-cobs, white ash, or beech.

The smoke-house is the best place in which to keep hams until they are wanted. If removed, they should be kept cool, dry, and free from flies. A canvas cover for each, saturated with lime, which may be put on with a whitewash brush, is a perfect protection against flies. When not to be kept long, they may be packed in dry salt, or even in sweet brine, without injury. A common method is to pack in dry oats, baked saw-dust, etc.

The following is the method in most general use in several

of the Western States. The chine is taken out, as also the spare-ribs from the shoulders, and the mouse-pieces and short-ribs, or griskins, from the middlings. No acute angles should be left to shoulders or hams. In salting up, all the meat, except the heads, joints, and chines, and smaller pieces, is put into powdering-tubs—water-tight half-hogsheads—or into large troughs, ten feet long and three or four feet wide at the top, made of the poplar tree. The latter are much more convenient for packing the meat in, and are easily caulked, if they should crack so as to leak. The salting-tray—or box in which the meat is to be salted, piece by piece, and from which each piece, as it is salted, is to be transferred to the powdering-tub, or trough—must be placed just so near the trough that the man standing between can transfer the pieces from one to the other easily, and without wasting the salt as they are lifted from the salting-box into the trough. The salter stands on the off-side of the salting-box. The hams should be salted first, the shoulders next, and the middlings last, which may be piled up two feet above the top of the trough or tub. The joints will thus in a short time be immersed in brine.

Measure into the salting-tray four measures of salt—a peck measure will be found most convenient—and one measure of clean, dry, sifted ashes; mix, and incorporate them well. The salter takes a ham into the tray, rubs the skin and the raw end with his composition, turns it over, and packs the composition of salt and ashes on the fleshy side till it is at least three-quarters of an inch deep all over it; and on the interior lower part of the ham, which is covered with the skin, as much as will lie on it. The man standing ready to transfer the pieces, deposits it carefully, without disturbing the composition, with the skin-side down, in the bottom of the trough. Each succeeding ham is then deposited, side by side, so as to leave the least possible space unoccupied.

When the bottom is wholly covered, see that every visible part of this layer of meat is covered with the composition of salt and ashes. Then begin another layer, every piece being covered on the upper or fleshy side three-quarters of an inch thick with the composition. When the trough is filled, even full, in this way, with the joints, salt the middlings with salt only, without the ashes, and pile them up on the joints so that the liquified salt may pass from them into the trough. Heads, joints, back bones, etc., receive salt only, and should not be put in the trough with the large pieces.

Much slighter salting will preserve them, if they are salted upon loose boards, so that the bloody brine from them can pass off. The joints and middlings are to remain in and above the trough without being re-handled, re-salted, or disturbed in any way, till they are to be hung up to be smoked.

If the hogs do not weigh more than one hundred and fifty pounds, the joints need not remain longer than five weeks in the pickle; if they weigh two hundred, or upward, six or seven weeks are not too long. It is better that they should stay in too long, rather than too short a time.

In three weeks, the joints, etc., may be hung up. Taking out of pickle, and preparing for hanging up to smoke, are thus performed: Scrape off the undissolved salt; if the directions have been followed, there will be a considerable quantity on all the pieces not immersed in the brine; this salt and the brine are all saved; the brine is boiled down, and the dry composition given to stock, especially to hogs. Wash every piece in lukewarm water, and with a rough towel clean off the salt and ashes. Next, put the strings in to hang up. Set the pieces up edgewise, that they may drain and dry. Every piece is then to be dipped into the meat-paint, as it is termed, composed of warm—not hot—water and very fine ashes, stirred together until they are of the consistence of thick paint, and hang up to smoke. By being thus dipped, they receive a coating which protects them from the fly, prevents dripping, and tends to lessen all external injurious influences. Hang up the pieces while yet moist with the paint, and smoke them well.

VALUE OF THE CARCASS.

No part of the hog is valueless, excepting, perhaps, the bristles of the fine-bred races. The very intestines are cleansed, and knotted into chittarlings, very much relished by some; the blood, mixed with fat and rice, is made into black puddings; and the tender muscle under the lumbar vertebræ is worked up into sausages, sweet, high-flavored, and delicious; the skin, roasted, is a rare and toothsome morsel; and a roast sucking-pig is a general delight; salt pork and bacon are in incessant demand, and form important articles of commerce.

One great value arises from the peculiarity of its fat, which, in contradistinction to that of the ox or of the sheep, is termed *lard*, and differs from either in the proportion of its constituent principles, which are essentially oleine and stearine. It is rendered, or fried out, in the same manner as mutton-suet. It melts completely at ninety-nine degrees Fahrenheit, and then has the appearance of a transparent and nearly colorless fixed oil. Eighty degrees is the melting-point. It consists of sixty-two parts oleine, and thirty-eight of stearine, out of one hundred When subjected to pressure between folds of blotting-paper, the oleine is absorbed, while the stearine remains. For domestic purposes, lard is much used; it is much better than butter for frying fish; and is much used in pastry, on the score of economy.

The stearine contains the stearic and margaric acids, which, when separated, are solid, and used as inferior substitutes for wax or spermaceti candles. The other, oleine, is fluid at a low temperature, and in American commerce is known as *lard-oil*, which is very pure, and extensively used for machinery, lamps, and most of the purposes for which olive or spermaceti oils are valued. It has given to pork a new and profitable use, by which the value of the carcass is greatly increased. A large amount of pork has thus been withdrawn from the market, and the depression, which must otherwise have occurred, has been thereby prevented.

Where the oil is required, the whole carcass, after taking out the hams and shoulders, is placed in a tub having two bottoms, the upper one perforated with holes. The pork is laid on the latter, and then tightly covered. Steam, at a high temperature, is then admitted into the tub, and in a short time all the fat is extracted, and falls upon the lower bottom. The remaining mass is bones and scraps. The last is fed to pigs, poultry, or dogs, or affords the best kind of manure. The bones are either used for manure, or are converted into animal charcoal, valuable for various purposes in the arts. When the object is to obtain lard of a fine quality, the animal is first skinned, and the adhering fat then carefully scraped off; thus avoiding the oily, viscid matter of the skin.

The *bristles* of the coarse breeds are long, strong, firm, and elastic. These are formed into brushes for painters and artists, as well as for numerous domestic uses. The *skin*, when tanned, is of a peculiar texture, and very tough. It is used for making pocket-books, and for some ornamental purposes; but chiefly for the seats of riding-saddles. The numerous little variegations on it, which constitute its beauty, are the orifices whence the bristles have been removed.

DISEASES AND THEIR REMEDIES

By reason of being generally considered a subordinate species of stock, swine do not, in many cases, share in the benefits which an improved system of agriculture and the present advanced state of veterinary science, have conferred upon other domesticated animals. Since they are by no means the most tractable of patients, it is any thing but an easy matter to compel them to swallow any thing to which their appetite does not incite them; and, hence, prevention will be found better than cure. *Cleanliness* is the great point to be insisted upon in the management of these animals If this, and warmth, be only attended to, ailments among them are comparatively rare.

As, however, disappointment may occasionally occur, even under the best system of management, a brief view of the principal complaints with which they are liable to be attacked is presented, together with the best mode of treatment to be adopted in such cases.

CATCHING THE PIG.

Swine are very difficult animals to obtain any mastery over, or to operate on, or examine. Seldom tame, or easily handled, they are at such periods most unmanageable—kicking, screaming, and even biting fiercely. The following method of getting hold of them has been recommended: Fasten a double cord to the end of a stick, and beneath the stick let there be a running noose in the cord; tie a piece of bread to the cord, and present it to the animal; and when he opens his mouth to seize the bait, catch the upper jaw in the noose. run it tight, and the animal is fast.

Another method is, to catch one foot in a running noose suspended from some place, so as to draw the imprisoned foot off the ground; or, to envelop the head of the animal in a cloth or sack.

All coercive measures, however, should, as far as possible, be avoided; for the pig is naturally so averse to being handled that in his struggles he will often do himself far more mischief than the disease which is to be investigated or remedied would effect.

BLEEDING.

The common mode of drawing blood from the pig is by cutting off portions of the ears or tail; this should only be resorted to when local and instant blood-letting is requisite. The jugular veins of swine lie too deep, and are too much imbedded in fat to admit of their being raised by any ligature about the neck; it is, therefore, useless to attempt to puncture them, as it would only be striking at random.

Those veins, however, which run over the interior surface of the ear, and especially toward its outer edge, may be opened without much difficulty; if the ear is turned back on the poll, one or more of them may easily be made sufficiently prominent to admit of its being punctured by pressing the fingers on the base of the ear, near to the conch. When the necessary quantity of blood has been obtained, the finger may be raised, and it will cease to flow.

The palate veins, running on either side of the roof of the mouth, are also easily opened by making two incisions, one on each side of the palate, about half way between the centre of the roof of the mouth and the teeth. The flow of blood may be readily stopped by means of a pledget of tow and a string, as in bleeding the horse.

The brachial vein of the fore-leg—commonly called the plate-vein—running along the inner side under the skin affords a good opportunity. The best place for puncturing it is about an inch above the knee, and scarcely half an inch backward from the radius, or the bone of the fore-arm. No danger need to be apprehended from cutting two or three times, if sufficient blood cannot be obtained at once. This vein will become easily discernible if a ligature is tied firmly around the leg, just below the shoulder.

This operation should always be performed with the lancet, if possible. In cases of urgent haste, where no lancet is at hand, a small penknife may be used; but the fleam is a dangerous and objectionable instrument.

DRENCHING.

Whenever it is possible, the medicine to be administered should be mingled with a portion of food, and the animal thus cheated or coaxed into taking it; since many instances are on record, in which the pig has ruptured some vessel in his struggles, and died on the spot, or so injured himself as to bring on inflammation and subsequent death.

Where this cannot be done, the following is the best method: Let a man get the head of the animal firmly between his knees—without, however, pinching it—while another secures the hinder parts. Then let the first take hold of the head from below, raise it a little, and incline it slightly toward the right, at the same time separating the lips on the left side so as to form a hole into which the fluid may be gradually poured—no more being introduced into the mouth at a time than can be swallowed at once. Should the animal snort or choke, the head must be released for a few moments, or he will be in danger of being strangled.

CATARRH.

This ailment—an inflammation of the mucous membranes of the nose, etc.—is, if taken in time, easily cured by opening medicines, followed up by warm bran-wash—a warm, dry sty—

and abstinence from rich grains, or stimulating, farinaceous diet. The cause, in most cases, is exposure to drafts of air, which should be guarded against.

CHOLERA.

For what is presented concerning this disease, the author is indebted to his friend, G. W. Bowler, V. S., of Cincinnati, Ohio, whose familiarity with the various diseases of our domestic animals and the best modes of treating them, entitles his opinions to great weight.

The term "cholera" is employed to designate a disease which has been very fatal among swine in different parts of the United States; and for the reason, that its symptoms, as well as the indications accompanying its termination, are very nearly allied to what is manifested in the disease of that name which visits man.

Epidemic cholera has, for several years past, committed fearful ravages among the swine of, particularly, Ohio, Indiana, and Kentucky. Indeed, many farmers who, until recently, have been accustomed to raise large numbers of these animals, are, in a great measure, disinclined to invest again in such stock, on account of the severe losses—in some instances to the extent of the entire drove upon particular places.

Various remedies have, of course, been prescribed; but the most have failed in nearly every case where the disease has secured a firm foothold. Preventives are, therefore, the most that can at present be expected; and in this direction something may be done. Although some peculiar change in the atmosphere is, probably, an impelling cause of cholera, its ravages may be somewhat stayed by removing other predisposing associate causes.

Granting that the hog is a filthy animal and fond of rooting among filth, it is by no means necessary to persist, for that reason, in surrounding him with all the nastiness possible; for even a hog, when penned up in a filthy place, in company with a large number of other hogs—particularly when that place is improperly ventilated—is not as healthy as when the animals are kept together in smaller numbers in a clean and well ventilated barn or pen. Look, for a moment, at a drove of hogs coming along the street, the animals all fat and ready for the knife. They have been driven several miles, and are scarcely able to crawl along, many of them having to be carried on drays, while others have died on the road. At last they are driven into a pen, perhaps, several inches deep with the manure and filth deposited there by hundreds of predecessors; every hole in the ground has become a puddle; and in such a place some one hundred or two hundred animals are piled together, exhausted from the drive which they have had. They lie down in the mud; and in a short time one can see the steam beginning to rise from their bodies in volumes, increasing their already prostrate condition by the consequent inhalation of the noxious gas thus thrown off from the system; the blood becomes impregnated with poison; the various functions of the body are thereby impaired; and disease will inevitably be developed in one form or another. Should the disease, known as the hog cholera, prevail in the neighborhood, the chances are very greatly in favor of their being attacked by it, and consequently perishing.

The *symptoms* of cholera are as follows: The animal

appears to be instantaneously deprived of energy; loss of appetite; lying down by himself; occasionally moving about slowly, as though experiencing some slight uneasiness internally; the eyes have a very dull and sunken appearance, which increases with the disease; the evacuations are almost continuous, of a dark color, having a fetid odor, and containing a large quantity of bile; the extremities are cold, and soreness is evinced when the abdomen is pressed; the pulse is quickened, and sometimes hardly perceptible, while the buccal membrane—that belonging to the cheek—presents a slight purple hue; the tongue has a furred appearance. The evacuations continue fluid until the animal expires, which may be in twelve hours from the first attack, or the disease may run on for several days.

In a very short time after death, the abdomen becomes of a dark purple color, and upon examination, the stomach is found to contain but a little fluid; the intestines are almost entirely empty, retaining a slight quantity of the dark colored matter before mentioned; the mucous membrane of the alimentary canal exhibits considerable inflammation, which sometimes appears only in patches, while the other parts are filled with dark venous blood—indicating a breaking up of the capillary vessels in such places.

Treatment. As a preventive, the following will be found valuable: Flour of sulphur, six pounds; animal charcoal, one pound; sulphate of iron, six ounces; cinchona pulverized, one pound. Mix well together in a large mortar; afterwards give a tablespoonful to each animal, mixed with a few potato-peelings and corn meal, three times a day. Continue this for one week, keeping the animal at the same time in a clean, dry place, and not allowing too many together.

CRACKINGS

These will sometimes appear on the skin of a hog, especially about the root of the ears and of the tail, and at the flanks. They are not at all to be confounded with mange, as they never result from any thing but exposure to extremes of temperature, while the animal is unable to avail himself of such protection as, in a state of nature, instinct would have induced him to adopt. They are peculiarly troublesome in the heat of summer, if he does not have access to water, in which to lave his parched limbs and half-scorched carcass.

Anoint the cracked parts twice or three times a day with tar and lard, well melted up together.

DIARRHŒA.

Before attempting to stop the discharge in this disease—which, it permitted to continue unchecked, will rapidly prostrate the animal, and probably terminate fatally—ascertain the quality ot food which the animal has recently had.

In a majority ot instances, this will be found to be the cause. If taken in its incipient stage, a mere change to a more binding diet, as corn, flour, etc., will suffice for a cure. If acidity is present—produced, probably, by the hog's having fed upon coarse, rank grasses in swampy places—give some chalk in the food, or powdered egg-shells, with about half a drachm of powdered rhubarb; the dose, of course, should vary with the size of the animal. In the acorn season, they alone will be found sufficiently curative, where facilities for obtaining them exist. Dry lodging is indispensable; and diligence is requisite to keep it dry and clean.

FEVER.

The *symptoms* of this disease are, redness of the eyes, dryness and heat of the nostrils, the lips, and the skin generally; appetite gone, or very defective; and, generally, a very violent thirst.

Bleed as soon as possible; after which house the animal well, taking care, at the same time, to have the sty well and

HUNTING THE WILD BOAR.

thoroughly ventilated. The bleeding will usually be followed, in an hour or two, by such a return of appetite as to induce the animal to eat a sufficient quantity of food to be made the vehicle for administering external remedies. The best is bread, steeped in broth. The hog, however, sinks so rapidly when his appetite is near gone, that no depletive medicines are, in general, necessary or proper; the fever will ordinarily yield to the bleeding, and the only object needs to be the support of his strength, by small portions of nourishing food, administered frequently.

Do not let the animal eat as much as his inclination might prompt; when he appears to be no longer ravenous, remove the mess, and do not offer it again until after a lapse of three or four hours. If the bowels are confined, castor and linseed oil, in equal quantities, should be added to the bread and broth, in the proportion of two to six ounces.

A species of fever frequently occurs as an *epizoötic*, oftentimes attacking the male pigs, and generally the most vigorous and best looking, without any distinction of age, and with a force and rapidity absolutely astonishing. At other times, its progress is much slower; the symptoms are less intense and alarming; and the veterinary surgeon, employed at the outset, may meet with some succ

The *causes* are, in the majority of instances, the bad styes in which the pigs are lodged, and the noisome food which they often contain. In addition to these is the constant lying on the dung-heap, whence is exhaled a vast quantity of deleterious gas; also, the remaining far too long on the muddy or parched ground, or too protracted exposure to the rigor of the season.

When an animal is attacked with this disease, he should be separated from the others, placed in a warm situation, some stimulating ointment applied to the chest, and a decoction of sorrel administered. Frictions of vinegar should also be applied to the dorsal and lumbar region. The drinks should be emolient, slightly imbued with nitre and vinegar, and with aromatic fumigation about the belly.

If the fever then appears to be losing ground, which may be

ascertained by the regularity of the pulse, by the absence of the plaintive cries before heard, by a less laborious respiration, by the absence of convulsions, and by the non-appearance of blotches on the skin, there is a fair chance of recovery. Then administer, every second hour, as before directed, and give a proper allowance of white water, with ground barley and rye.

When the symptoms redouble in intensity, it is best to destroy the animal; for it is rare that, after a certain period, much chance of recovery exists. Bleeding is seldom of much avail, but produces, occasionally, considerable loss of vital power, and augments the putrid diathesis.

FOUL SKIN.

A simple irritability or foulness of skin will usually yield to cleanliness, and a washing with a solution of chloride of lime; but, if it is neglected for any length of time, it assumes a malignant character—scabs and blotches, or red and fiery eruptions appear—and the disease rapidly passes into mange, which will be hereafter noticed.

INFLAMMATION OF THE LUNGS.

This disease, popularly known as heavings, is scarcely to be regarded as curable. Were it observed in its first stage, when indicated by loss of appetite and a short, hard cough, it might, possibly, be got under by copious bleeding, and friction with stimulating ointment on the region of the lungs; minute and frequent doses of tartar emetic should also be given in butter—all food of a stimulating nature carefully avoided—and the animal kept dry and warm. If once the heavings set in, it may be calculated with confidence that the formation of tubercles in the substance of the lungs has begun; and when these are formed, they are very rarely absorbed.

The *causes* of the disease are damp lodging, foul air, want of ventilation, and unwholesome food. When tubercular formation becomes established, the disease may be communicated through the medium of the atmosphere, the infectious influence depending upon the noxious particles respired from the lungs of the diseased animal.

The following may be tried, though the knife is probably the best resort, if for no other reason, at least to provide against the danger of infection: Shave the hair away from the chest, and beneath each fore-leg; wet the part with spirits of turpentine, and set fire to it, having previously had the animal well secured, with his head well raised, and a flannel cloth at hand with which to extinguish the flame after it has burned a sufficient time to produce slight blisters; if carried too far, a sore is formed, productive of no good effects, and causing unnecessary suffering. Calomel may also be used, with a view to promote the absorption of the tubercles; but the success is questionable.

JAUNDICE.

The *symptoms* of this disease are, yellowness of the white of the eye; a similar hue extending to the lips; and sometimes, but not invariably, swelling of the under part of the jaw.

Treatment. Bleed freely; diminish the quantity of food; and give an active aperient every second day. Aloes are, perhaps, the best, combined with colocynth; the dose will vary with the size of the animal.

LEPROSY.

This complaint commonly commences with the formation of a small tumor in the eye, followed by a general prostration of spirits; the head is held down; the whole frame inclines toward the ground; universal languor succeeds; the animal refuses food, languishes, and rapidly falls away in flesh; blisters soon make their appearance beneath the tongue, then upon the throat, the jaws, the head, and the entire body.

The *Causes* of this disease are want of cleanliness, absence of fresh air, want of due attention to ventilation, and foul feeding. The obvious *treatment*, therefore, is, first, bleed; clean out the sty daily; wash the affected animal thoroughly with soap and water, to which soda or potash has been added; supply him with a clean bed; keep him dry and comfortable; let him have gentle exercise, and plenty of fresh air; limit the quantity of his food, and diminish its rankness; give bran with wash, in which add, for an average-sized hog—say one of one hundred and sixty pounds weight—a tablespoonful of the flour of sulphur, with as much nitre as will cover a dime, daily. A few grains of powdered antimony may also be given with effect.

LETHARGY.

Symptoms: torpor; desire to sleep; hanging of the head; and, frequently, redness of the eyes. The origin of this disease is, apparently, the same as that of indigestion, or surfeit, except that, in this instance, it acts upon a hog having a natural tendency to a redundancy of blood.

Treatment. Bleed copiously; then administer an emetic. A decoction of camomile flowers will be safest; though a sufficient dose of tartar emetic will be far more certain. After this, reduce for a few days the amount of the animal's food, and administer a small portion of nitre and sulphur in each morning's meal.

MANGE.

This cutaneous affection owes its existence to the presence of a minute insect, called *acarus scabiei*, or mange-fly, which burrows beneath the cuticle, and occasions much irritation and annoyance in its progress through the skin.

Its *symptoms* are sufficiently well known, consisting of scabs, blotches, and sometimes multitudes of minute pustules on different parts of the body. If neglected, these symptoms become aggravated; the disease spreads rapidly over the entire surface of the skin, and if allowed to proceed on its course unchecked, will before long produce deep-seated ulcers and malignant sores, until the whole carcass of the affected animal becomes a mass of corruption.

The *cause* is to be looked for in dirt, accompanied by hot-feeding. Hogs, however well and properly kept, will occasionally become affected with this disease from contagion. Few diseases are more easily propagated by contact than mange. The introduction of a single affected pig into an establishment may, in one night, cause the seizure of scores of others. No foul-skinned pigs, therefore, should be introduced into the piggery; indeed, it would be an excellent precaution to wash every animal newly purchased with a strong solution of chloride of lime.

Treatment. If the mange is but of moderate violence, and not of very long standing, the best mode is to wash the animal, from snout to tail, leaving no portion of the body uncleansed, with soft soap and water. Place him in a dry and clean sty, which is so situated as to command a constant supply of fresh air, without, at the same time, an exposure to cold or draught; furnish a bed of clean, fresh straw. Reduce his food, both in quality and quantity; let boiled or steamed roots, with buttermilk, or dairy-wash take the place of any food of a heating or inflammatory character. Keep him without food for five or six hours, and then give to a hog of average size two ounces of Epsom salts in a warm bran mash—to be increased or diminished, of course, as the animal's size may require. This should be previously mixed with a pint of warm water, and added to about half a gallon of warm bran mash, and it will act as a gentle purgative. Give in every meal afterward one tablespoonful of flour of sulphur, and as much nitre as will cover a dime, for from three days to a week, according to the state of the disease. When the scabs begin to heal, the pustules to retreat, and the fiery sores to fade, a cure may be anticipated.

When the above treatment has been practised for fourteen days, without effecting a cure, prepare the following: train oil, one pint; oil of tar, two drachms; spirits of turpentine, two drachms; naphtha, one drachm; with as much flour of sulphur as will form the foregoing into a thick paste. Rub the animal previously washed with this mixture; let no portion of the hide escape. Keep the hog dry and warm after this application, and allow it to remain on his skin for three days. On the fourth day wash him again with soft soap, adding a small quantity of soda to the water. Dry him well afterward, and let him remain as he is, having again changed his bedding, for a day or so; continue the sulphur and nitre as before. Almost all cases of mange, however obstinate, will, sooner or later, yield to this treatment. After he is convalescent, whitewash the sty, and fumigate it by placing a little chloride of lime in a cup, or other vessel, and pouring a little vitriol upon it. In the absence of vitriol, boiling water will answer nearly as well.

MEASLES.

This is one of the most common diseases to which hogs are liable. The *symptoms* are, redness of the eyes, foulness of the skin, and depression of spirits; decline, or total departure of the appetite; small pustules about the throat, and red and purple eruptions on the skin. The last are more plainly visible after death, when they impart a peculiar appearance to the grain of the meat, with fading of its color, and distention of the fibre, giving an appearance similar to that which might be produced by puncturing the flesh.

Treatment. Allow the animal to fast, in the first instance, for twenty-four hours, and then administer a warm drink, containing a drachm of carbonate of soda, and an ounce of bole armenian; wash the animal, cleanse the sty, and change the bedding; give at every feeding, or thrice a day, thirty grains of flour of sulphur, and ten of nitre.

This malady is attributable to dirt, combined with the giving of steamed food or wash to hogs at too high a temperature. It is troublesome to eradicate, but usually yields to treatment, and is rarely fatal.

MURRAIN.

This resembles leprosy in its *symptoms*, with the addition of staggering, shortness of breath, and discharge of viscid matter from the eyes and mouth.

The *treatment* should consist of cleanliness, coolness, bleeding, purging, and limitation of food. Cloves of garlic are recommended; and as in all febrile diseases there exists a greater or less disposition to putrefaction, it is probable that garlic, from its antiseptic properties, may be useful.

QUINSY.

This is an inflammatory affection of the glands of the throat.

Treatment. Shave away the hair, and rub with tartar-emetic ointment. Fomenting with very warm water is also useful. When external suppuration takes place, it is to be regarded as a favorable symptom. In this case, wait until the swellings are thoroughly ripe; then with a sharp knife make an incision through the entire length, press out the matter, wash with warm water, and afterward dress the wound with any resinous ointment, or yellow soap with coarse brown sugar.

STAGGERS.

This disease is caused by an excessive determination of blood to the head.

Treatment. Bleed freely and purge.

SWELLING OF THE SPLEEN.

The *symptom* most positively indicative of this disease is the circumstance of the affected animal leaning toward one side, cringing, as it were, from internal pain, an bending toward the ground.

The *cause* of the obstruction on which the disease depends, is over-feeding—permitting the animal to indulge its appetite to the utmost extent that gluttony may prompt, and the capacity of its stomach admits. A very short perseverance in this mode of management—or, rather, mismanagement—will produce this, as well as other maladies, deriving their origin from a depraved condition of the secretions and the obstruction of the excretory ducts.

Treatment. Clean out the alimentary canal by means of a powerful aperient. Allow the animal to fast for four or five hours, when he will take a little sweet wash or broth, in which may be mingled a dose of Epsom salts proportioned to his size. This will generally effect the desired end—a copious evacuation—and the action of the medicine on the watery secretions will also relieve the existing diseased condition of the spleen.

If the affection has continued for any length, the animal should be bled. A decoction of the leaves and tops of wormwood and liverwort, produced by boiling them in soft water for six hours, may be given in doses of from half a pint to a pint and a half, according to the size, age, etc., of the animal. Scammony and rhubarb, mixed in a bran wash, or with Indian meal, may be given with advantage on the following day; or, equal portions of blue-pill mass and compound colocynth pill, formed into a bolus with butter. The animal having been kept fasting the previous night, will probably swallow it; if not, let his fast continue a couple of hours longer. Lower his diet, and keep him on reduced fare, with exercise, and, if it can be managed, grazing, until the malady has passed away. If he is then to be fattened, it should be done gradually; be cautious of at once restoring him to full diet.

SURFEIT.

This is another name for indigestion. The *symptoms* are, panting; loss of appetite; swelling of the region about the stomach, etc.; and frequently throwing up the contents of the stomach.

Treatment. In general, this affection will pass away, provided only it is allowed to cure itself, and all food carefully kept from the animal for a few hours; a small quantity of sweet grains, with a little bran mash, may then be given, but not nearly as much as the animal would wish to take. For a few days, the food should be limited in quantity, and of a washy, liquid nature. The ordinary food may then be resumed, only observing to feed regularly, and remove the fragments remaining after each meal.

TUMORS.

These are hard swellings, which make their appearance on different parts of the body. They are not formidable, and require only to be suffered to progress until they soften; then make a free incision, and press out the matter. Sulphur and nitre should be given in the food, as the appearance of these swellings, whatever be their cause. indicates the necessity of alterative medicines.

THE DOMESTIC FOWL. The cock tribe is used as a generic term, to include the whole family of domestic fowls; the name of the male, in this instance, furnishing an appellation sufficiently comprehensive and well recognized.

The domestic cock appears to have been known to man from a very early period. Of his real origin there is little definitely known; and even the time and manner of his introduction into Greece, or Southern Europe, are enveloped in obscurity. In the palmiest days of Greece and Rome, however, he occupied a conspicuous place in those public shows which amused the masses of the people. He was dedicated to the service of the pagan deities, and was connected with the worship of Apollo, Mercury, Mars, and particularly Esculapius. The flesh of this bird was highly esteemed as a delicacy, and occupied a prominent place at the Roman banquets. Great pains were taken in the rearing and fattening of poultry for this purpose.

The practice of cock-fighting, barbarous as it is, originated in classic times, and among the most polished and civilized people of antiquity. To its introduction into Britain by the Cæsars we owe our acquaintance with the domestic fowl.

It is impossible to state positively to what species of the wild cock, known at present, we are to look for the primitive type, so remote is the date of the original domestication of the fowl. Many writers have endeavored to show that all the varieties of the domestic fowl, of which we now have knowledge, are derived from a single primitive stock. It has, also. been confidently asserted that the domestic cock owes his origin to the jungle iowl of India. The most probable supposition, however, is, that the varieties known to us may be referred to a few of the more remarkable fowls, as the progenitors of the several species. The great fowl of St. Jago and Sumatra may, perhaps, safely be recognized as the type of some of the larger varieties, such as the Spanish and the Padua fowls, and those resembling them; while to the Bankiva cock, probably, the smaller varieties belong, such as Bantams, the Turkish fowl, and the like.

The reasons assigned for supposing these kinds to be the true originals of our domestic poultry, are, *first*, the close resemblance subsisting between their females and our domestic hens; *second*, the size of our domestic cock being intermediate between the two, and alternating in degree, sometimes inclining toward the one, and sometimes toward the other; *third*, from the nature of their feathers and their general aspect—the form and distribution of their tails being the same as our domestic fowls; and, *fourth*, in these two birds alone are the females provided with a crest and small wattles, characteristics not to be met with in any other wild species.

The wild cock, or the St. Jago fowl, is frequently so tall as to be able to peck crumbs without difficulty from an ordinary dinner-table. The weight is usually from ten to thirteen or fourteen pounds. The comb of both cock and hen is large, crown-shaped, often double, and sometimes, but not invariably, with a tufted crest of feathers, which occurs with the greatest frequency, and grows to the largest size, in the hen. The voice is strong and very harsh; and the young do not arrive to full plumage until more than half grown.

The Bankiva fowl is a native of Java, and is characterized by a red indented comb, red wattles, and ashy-gray legs and feet. The comb of the cock is scolloped, and the tail elevated a little above the rump, the feathers being disposed in the form of tiles or slates; the neck-feathers are of a gold color, long, dependent, and rounded at the tips; the head and neck are of a fawn color; the wing coverts a dusky brown and black; the tail and belly, black. The color of the hen is a dusky ash-gray and yellow; her comb and wattles much smaller than those of the cock, and—with the exception of the long hackles—she has no feathers on her neck. These fowl are exceedingly wild, and inhabit the skirts of woods, forests, and other savage and unfrequented places These Bankivas resemble our Bantams very much; and, like them, are also occasionally to be seen feathered to the feet and toes.

Independent of all considerations of profitableness, domestic fowls are gifted with two qualifications, which—whether in man, beast, or bird—are sure to be popular: a courageous temper and an affectionate disposition. When we add to these beauty of appearance and hardiness of constitution, it is no wonder that they are held in such universal esteem.

The courage of the cock is emblematic, his gallantry admirable, and his sense of discipline and subordination most exemplary. The hen is deservedly the acknowledged pattern of maternal love. When her passion of philoprogenitiveness is disappointed by the failure or subtraction of her own brood, she will either continue incubating till her natural powers fail, or will violently kidnap the young of other fowls, and insist upon adopting them.

It would be idle to attempt an enumeration here of the numerous breeds and varieties of the domestic fowl. Those only, therefore, will be described which are generally accepted as the best varieties; and these arranged, not in the order of their merits necessarily, but alphabetically, for convenience of reference.

THE BANTAM.

The original of the Bantam is, as has been already remarked, the Bankiva fowl. The small white, and also the colored Bantams, whose legs are heavily feathered, are sufficiently well-known to render a particular description unnecessary. Bantam-fanciers generally prefer those which have clean, bright legs, without any vestige of feathers. A thorough-bred cock, in their judgment, should have a rose comb; a well-feathered tail, but without the sickle feathers; full hackles; a proud, lively carriage; and ought not to exceed a pound in weight. The nankeen-colored, and the black are the general favorites.

THE BANTAM.

These little creatures exhibit some peculiar habits and traits of disposition. Amongst others, the cocks are so fond of sucking the eggs laid by the hen that they will often drive her from the nest in order to obtain them; they have even been known to attack her, tear open the ovarium, and devour its shell-less contents. To prevent this, first a hard-boiled, and then a marble egg may be given them to fight with, taking care, at the same time, to prevent their access either to the hen or to any real eggs. Another strange propensity is a passion for sucking each other's blood, which is chiefly exhibited when they are moulting, when they have been known to peck each other naked, by pulling out the new feathers as they appear, and squeezing with their beaks the blood from the bulbs at the base. These fowls being subject to a great heat of the skin, its surface occasionally becomes hard and tightened; in which cases the hard roots of the feathers are drawn into a position more nearly at right angles with the body than at ordinary times, and the skin and superficial muscles are thus subjected to an unusual degree of painful irritation. The disagreeable habit is, therefore, simply a provision of Nature for their relief, which may be successfully accomplished by washing with warm water, and the subsequent application of pomatum to the skin.

BANTAM.

Bantams, in general, are greedy devourers of some of the most destructive of our insects; the grub of the cock-chafer and the crane-fly being especial favorites with them. Their chickens can hardly be raised so well, as by allowing them free access to minute insect dainties; hence, the suitableness of a worn-out hotbed for them during the first month or six weeks. They are thus positively serviceable creatures to the farmer, as far as their limited range extends; and still more so to the gardener and the nurseryman, as they will save various garden crops from injuries to which they would otherwise be exposed.

The fowl commonly known as the Bantam is a small, elegantly-formed, and handsomely tinted variety, evidently but remotely allied to the game breed, and furnished with feathers to the toes.

The African Bantam. The cock of this variety is red upon the neck, back, and hackles; tail, black and erect, studded with glossy green feathers upon the sides; breast, black ground spotted with yellow, like the Golden Pheasant; comb, single; cheeks, white or silvery; the pullet is entirely black, except the inside of the wing-tips, which is perfectly white. In size, they compare with the common pigeon, being very small; their wings are about two inches longer than their bodies; and their legs dark and destitute of feathers. They are very quiet, and of decided benefit in gardens, in destroying bugs.

These symmetrically-formed birds are highly prized, both by the fancier and the practical man, and the pure-bloods are very rare. They weigh from eight to twelve ounces each for the hens; and the cocks, from sixteen to twenty ounces.

THE BOLTON GRAY.

These fowls—called, also, Dutch Every-day Layers, Pencilled Dutch Fowl, Chittaprats, and, in Pennsylvania, Creole

BOLTON GRAYS OR CREOLE FOWL.

Fowl—were originally imported from Holland to Bolton, a town in Lancashire, England, whence they were named.

They are small sized, short in the leg, and plump in the make; color of the genuine kind, invariably pure white in the whole cappel of the neck; the body white, thickly spotted with black, sometimes running into a grizzle, with one or more black bars at the extremity of the tail. A good cock of this breed may weigh from four to four and a half pounds; and a hen from three to three and a half pounds.

The superiority of a hen of this breed does not consist so much in rapid as in continued laying. She may not produce as many eggs in a month as some other kinds, but she will, it is claimed, lay more months in the year than probably any other variety. They are said to be very hardy; but their eggs, in the judgment of some, are rather watery and innutritious.

THE BLUE DUN.

The variety known under this name originated in Dorsetshire, England. They are under the average size, rather slenderly made, of a soft and pleasing bluish-dun color, the neck being darker, with high, single combs, deeply serrated. The cock is of the same color as the hen, but has, in addition, some handsome dark stripes in the long feathers of the tail, and sometimes a few golden, or even scarlet marks, on the wings. They are exceedingly impudent, familiar, and pugnacious.

The hens are good layers, wanting to sit after laying a moderate number of eggs, and proving attentive and careful rearers of their own chickens, but rather savage to those of other hens. The eggs are small and short, tapering slightly at one end, and perfectly white. The chickens. on first coming from the egg, sometimes bear a resemblance to the gray and yellow catkin of the willow, being of a soft bluish gray, mixed with a little yellow here and there.

Some class these birds among the game fowls, not recognizing them as a distinct race, upon the ground that, as there are Blue Dun families belonging to several breeds—the Spanish, the Polish, the Game, and the Hamburghs, for example—it is more correct to refer each Blue Dun to its own proper ancestry.

THE CHITTAGONG.

The Chittagong is a very superior bird, showy in plumage, exceedingly hardy, and of various colors. In some, the gray predominates, interspersed with lightish yellow and white feathers upon the pullets. The legs are of a reddish flesh-color; the meat is delicately white, the comb large and single, wattles very full, wings good size. The legs are more or less feathered; the model is graceful, carriage proud and easy, and action prompt and determined.

This breed is the largest in the world; the pullets usually weighing from eight to nine pounds when they begin to lay, and the cocks from nine to ten pounds at the same age. They do not lay as many eggs in a year as smaller hens; but they lay as many pounds of eggs as the best breeds. This breed has been, by some, confounded with the great Malay; but the points of difference are very noticeable. There is less offal; the flesh is finer, although the size is greatly increased; their fecundity is greater; and the offspring arrive earlier at maturity than in the common Malay variety.

There is also a *red* variety of the Chittagong, which is

rather smaller than the gray. These have legs sometimes yellow and sometimes blue; the latter color, perhaps, from some mixture with the dark variety; the wings and tail are short. Sometimes there is a rose-colored comb, and a top-knot, through crossing. This variety may weigh sixteen or eighteen pounds a pair, as ordinarily bred. The eggs are large and rich, but not very abundant, and they do not hatch remarkably well.

There is, besides, a *dark-red* variety; the hens yellow or brown, with single serrated comb, and no top-knot; legs heavily feathered, the feathers black and the legs yellow The cock is black on the breast and thighs.

The Chittagongs are generally quite leggy, standing some twenty-six inches high; and the hens twenty-two inches. A first cross with the Shanghae makes a very large and valuable bird for the table, but not for breeding purposes.

THE COCHIN CHINA.

The Cochin China fowl are said to have been presented to Queen Victoria from the East Indies. In order to promote their propagation, her majesty made presents of them occasionally to such persons as she supposed likely to appreciate them. They differ very little in their qualities, habits, and general appearance from the Shanghaes, to which they are undoubtedly nearly related. The egg is nearly the same size, shape, and color; both have an equal development of comb and wattles—the Cochins slightly differing from the Shanghaes, chiefly in being somewhat fuller and deeper in the breast, not quite so deep in the quarter, and being usually smooth-legged, while the Shanghaes, generally, are more or less heavily feathered. The plumage is much the same in both cases; and the crow in both is equally sonorous and prolonged, differing considerably from that of the Great Malay.

The cock has a large, upright, single, deeply-indented comb,

COCHIN CHINAS.

very much resembling that of the Black Spanish, and, when in high condition, of quite as brilliant a scarlet; like him, also, he has sometimes a very large white ear-hole on each cheek, which, if not an indispensable or even a required qualification, is, however, to be preferred, for beauty at least. The wattles are large, wide, and pendent. The legs are of a pale flesh-color; some specimens have them yellow, which is objectionable. The feathers on the breast and sides are of a bright chestnut-brown, large and well-defined, giving a scaly or imbricated appearance to those parts. The hackle of the neck is of a light yellowish brown; the lower feathers being tipped with dark brown, so as to give a spotted appearance to the neck. The tail-feathers are black, and darkly iridescent; back, scarlet-orange; back-hackle, yellow-orange. It is, in short, altogether a flame-colored bird. Both sexes are lower in the leg than either the Black Spanish or the Malay.

The hen approaches in her build more nearly to the Dorking than to any other breed, except that the tail is very small and proportionately depressed; it is smaller and more horizontal than in any other fowl. Her comb is of moderate size, almost small; she has, also, a small, white ear-hole. Her coloring is flat, being composed of various shades of very light brown, with light yellow on the neck. Her appearance is quiet, and only attracts attention by its extreme neatness, cleanliness, and compactness.

The eggs average about two ounces each They are smooth, of an oval shape, equally rounded at each end, and of a rich buff color, nearly resembling those of the Silver Pheasant. The newly-hatched chickens appear very large in proportion to the size of the egg. They have light, flesh-colored bills, feet, and legs, and are thickly covered with down, of the hue commonly called "carroty." They are not less thrifty than any other chickens, and feather somewhat more uniformly than either the Black Spanish or the Malay. It is, however, most desirable to hatch these—as well as other large-growing varieties—as early in the spring as possible; even so soon as the end of February. A peculiarity in the cockerels is, that they do not show even the rudiments of their tail-feathers till they are nearly full-grown. They increase so rapidly in other directions, that there is no material to spare for the production of these decorative appendages.

The merits of this breed are such that it may safely be recommended to people residing in the country. For the inhabitants of towns it is less desirable, as the light tone of its plumage would show every mark of dirt and defilement; and the readiness with which they sit would be an inconvenience, rather than otherwise, in families with whom perpetual layers are most in requisition. Expense apart, they are equal or superior to any other fowl for the table; their flesh is delicate, white, tender, and well flavored.

THE CUCKOO.

The fowl so termed in Norfolk, England, is, very probably, an old and distinct variety; although they are generally regarded as mere Barn-door fowls—that is, the merely accidental result of promiscuous crossing.

The name probably originated from its barred plumage, which resembles that on the breast of the Cuckoo. The prevailing color is a slaty blue, undulated, and softly shaded with white all over the body, forming bands of various widths. The comb is very small; irides, bright orange; feet and legs, light flesh-color. The hens are of a good size; the cocks are large, approaching the heaviest breeds in weight. The chickens, at two or three months old, exhibits the barred plumage even more perfectly than the full-grown birds. The eggs average about two ounces each, are white, and of porcelain smoothness. The newly-hatched chickens are gray, much resembling those of the Silver Polands, except in the color of the feet and legs.

This breed supplies an unfailing troop of good layers, good sitters, good mothers and good feeders; and is well worth promotion in the poultry-yard.

THE DOMINIQUE.

This seems to be a tolerably distinct and permanent variety, about the size of the common Dunghill Fowl. Their combs are generally double—or rose, as it is sometimes called—and the wattles small. Their plumage presents, all over, a sort of greenish appearance, from a peculiar arrangement of blue and white feathers, which is the chief characteristic of the variety; although, in some specimens, the plumage is inevitably gray in both cock and hen. They are very hardy, healthy, excellent layers, and capital incubators. No fowl have better stood the tests of mixing without deteriorating than the pure Dominique.

Their name is taken from the island of Dominica, from which they are reported to have been imported. Take all in all, they are one of the very best breeds of fowl which we have; and although they do not come in to laying so young as the Spanish, they are far better sitters and nursers.

THE DORKING.

This has been termed the Capon Fowl of England. It forms the chief supply for the London market, and is distinguished by a white or flesh-colored smooth leg, armed with five, instead of four toes, on each foot. Its flesh is extremely delicate, especially after caponization; and it has the advantage over some other fowls of feeding rapidly, and growing to a very respectable size when properly managed.

For those who wish to stock their poultry yards with fowls of the most desirable shape and size, clothed in rich and varigated plumage, and, not expecting perfection, are willing to overlook one or two other points, the Speckled Dorkings —so called from the town of Surrey, England, which brought them into modern repute—should be selected. The hens, in addition to their gay colors, have a large, vertically flat comb, which, when they are in high health, adds very much to their brilliant ap-

WHITE DORKINGS.

pearance, particularly if seen in bright sunshine. The cocks are magnificent. The most gorgeous hues are lavished upon them, which their great size and peculiarly square-built form display to the greatest advantage. Their legs are short; their breast broad; there is but a small proportion of offal; and the good, profitable flesh is abundant. The cocks may be brought to considerable weight, and the flavor and appearance of their meat are inferior to none. The eggs are produced in reasonable abundance; and, though not equal in size to those of Spanish hens, may fairly be called large.

They are not everlasting layers, but at due or convenient

intervals manifest the desire of sitting. In this respect, they are steady and good mothers when the little ones appear. They are better adapted than any other fowl, except the Malay, to hatch superabundant turkeys' eggs; as their size and bulk enable them to afford warmth and shelter to the young for a long period. For the same reason, spare goose eggs may be entrusted to them.

With all these merits, however, they are not found to be a profitable breed, if kept thorough-bred and unmixed. Their powers seem to fail at an early age. They are also apt to pine away and die just at the point of reaching maturity. They appear at a certain epoch to be seized with consumption —in the Speckled Dorkings, the lungs seeming to be the seat of the disease. The White Dorkings are, however, hardy and active birds, and are not subject to consumption or any other disease.

As mothers, an objection to the Dorkings is, that they are too heavy and clumsy to rear the chickens of any smaller and more delicate bird than themselves. Pheasants, partridges, bantams, and Guinea fowl are trampled under foot and crushed, if in the least weakly. The hen, in her affectionate industry in scratching for grub, kicks her smallest nurslings right and left, and leaves them sprawling on their backs; and before they are a month old, half of them will be muddled to death with this rough-kindness.

In spite of these drawbacks, the Dorkings are still in high favor; but a cross is found to be more profitable than the true breed. A glossy, energetic game-cock, with Dorking hens, produces chickens in size and beauty little inferior to their maternal parentage, and much more robust. The supernumerary toe on each foot almost always disappears with the first cross; but it is a point which can very well be spared without much disadvantage. In other respects, the appearance of the newly-hatched chickens is scarcely altered. The eggs of the Dorkings are large, pure white, very much rounded, and nearly equal in size at each end. The chickens are brownish-yellow, with a broad brown stripe down the middle of the back, and a narrower one on each side; feet and legs yellow.

THE FAWN-COLORED DORKING. The fowl bearing this name is a cross between the white Dorking and the fawn-colored Turkish fowl. They are of lofty carriage, handsome, and healthy. The males of this breed weigh from eight to nine pounds, and the females from six to seven; and they come to maturity early for so large a fowl. Their tails are shorter and their eggs darker than those of other Dorkings; their flesh is fine and their eggs rich. It is one of the best varieties of fowl known, as the size is readily increased without diminishing the fineness of the flesh.

THE BLACK DORKING. The bodies of this variety are of a large size, with the usual proportions of the race, and of a jet-black color. The neck-feathers of some of the cocks are tinged with a bright gold color, and those of some of the hens bear a silvery complexion. Their combs are usually double, and very short, though sometimes cupped, rose, or single, with wattles small; and they are usually very red about the head. Their tails are rather shorter and broader than most of the race, and they feather rather slowly. Their legs are short and black, with five toes on each foot, the bottom of which is sometimes yellow. The two back toes are very distinct, starting from the foot separately; and there is frequently a part of an extra toe between the two.

This breed commence laying when very young, and are very thrifty layers during winter. Their eggs are of a large size, and hatch well; they are perfectly hardy, as their color indicates, and for the product are considered among the most valuable of the Dorking breed.

THE DUNGHILL FOWL.

This is sometimes called the Barn-door fowl, and is characterized by a thin, serrated, upright comb, and wattles hanging from each side of the lower mandible; the tail rises in an arch, above the level of the rump; the feathers of the rump are long and line-like; and the color is finely variegated. The female's comb and wattles are smaller than those of the cock; she is less in size, and her colors are more dull and sombre.

In the best specimens of this variety, the legs should be white and smooth, like those of the Dorking, and their bodies round and plump. Being mongrels, they breed all colors, and are usually from five to seven or eight pounds per pair.

THE FRIZZLED FOWL.

This fowl is erroneously supposed to be a native of Japan, and, by an equally common error, is frequently called the "Friesland," under the apprehension that it is derived from that place. Its name, however, originates from its peculiar appearance. It is difficult to say whether this is an aboriginal variety, or merely a peculiar instance of the morphology of feathers; the circumstance that there are also frizzled Bantams, would seem to make in favor of the latter position.

The feathers are ruffled or frizzled, and the reversion makes them peculiarly susceptible of cold and wet, since their plumage is of little use as clothing. They have thus the demerit of being tender as well as ugly. In good specimens, every feather looks as if it had been curled the wrong way with a pair of hot curling-irons. The plumage is variegated in its colors; and there are two varieties, called the Black and White Frizzled. The stock, which is rather curious than valuable, is retained in this country more by importation than by rearing.

Some writers say that this variety is a native of Asia, and that it exists in a domestic state throughout Java, Sumatra, and all the Philippine islands, where it succeeds well. It is, according to such, uncertain in what country it is still found wild.

THE GAME FOWL.

It is probable that these fowl, like other choice varieties, are natives of India. It is certain that in that country an original race of some fowl exists, at the present day, bearing in full perfection all the peculiar characteristics of the species. In India, as is well known, the natives are infected with a passion for cock-fighting. These fowls are carefully bred for this barbarous amusement, and the finest birds become articles of great value. In Sumatra, the inhabitants are so much addicted to the cruel sports to which these fowls are devoted, that instances are recorded of men staking not only their property upon the issue of a fight, but even their wives and children. The Chinese are likewise passionately fond of this pastime; as, indeed, are all the inhabitants of the Indian countries professing the Mussulman creed. The Romans introduced the practice into Britain, in which country the earliest recorded cock-fight dates back to about the year 1100. In Mexico and the South American countries it is still a national amusement.

GRAY GAME FOWLS.

The game fowl is one of the most gracefully formed and beautifully colored of our domestic breeds of poultry; and in its form, aspect, and that extraordinary courage which characterizes its natural disposition, exhibits all that either the naturalist or the sportsman would at once recognize as the purest type of high blood, embodying, in short, all the most indubitable characteristics of gallinaceous aristocracy.

It is somewhat inferior in size to other breeds, and in its shape approximates more closely to the elegance and lightness of form usually characteristic of a pure and uncontaminated race. Amongst poultry, he is what the Arabian is amongst horses, the high-bred Short-horn amongst cattle, and the fleet greyhound amongst the canine race.

The flesh is beautifully white, as well as tender and delicate. The hens are excellent layers, and although the eggs are under the average size, they are not to be surpassed in excellence of flavor. Such being the character of this variety of fowl, it would doubtless be much more extensively cultivated than it is, were it not for the difficulty attending the rearing of the young; their pugnacity being such, that a brood is scarcely feathered before at least one-half are killed or blinded by fighting.

With proper care, however, most of the difficulties to be apprehended may be avoided. It is exceedingly desirable to perpetuate the race, for uses the most important and valuable. As a cross with other breeds, they are invaluable in improving the flavor of the flesh, which is an invariable consequence. The plumage of all fowl related to them is increased in brilliancy; and they are, moreover, very prolific, and the eggs are always enriched.

THE MEXICAN HEN-COCK. This unique breed is a favorite variety with the Mexicans, who term them Hen-cocks from the fact that the male birds have short, broad tails, and, in color and plumage, the appearance of the hens of the same variety, differing only in the comb, which is very large and erect in the cock, and small in the hen. They are generally pheasant-colored, with occasional changes in plumage from a light yellow to a dark gray; and, in some instances, there is a tendency to black tail-feathers and breast, as well as an inclination to gray and light yellow, and with a slight approximation to red hackles in some rare instances.

This variety has a strong frame, and very large and muscular thighs. The cocks are distinguished by large, upright combs, strong bills, and very large, lustrous eyes. The legs vary from a dirty to a dark-green color. The hen does not materially differ in appearance from the cock. They are as good layers and sitters as any other game breed, and are good nurses.

THE WILD INDIAN GAME. This variety was originally imported into this country from Calcutta. The hen has a long neck, like a wild goose; neither comb nor wattles; of a dark, glossy green color; very short fan tail; lofty in carriage, trim built, and wild in general appearance; legs very large and long, spotted with blue; ordinary weight from four and a half to six pounds. As a layer, she is equal to other fowls of the game variety.

The cock stands as high as a large turkey, and weighs nine pounds and upward; the plumage is of a reddish cast, interspersed with spots of glossy green; comb very small; no wattles; and bill unlike every other fowl, except the hen.

THE SPANISH GAME. This variety is called the English fowl by some writers. It is more slender in the body, the neck, the bill, and the legs, than the other varieties, and the colors, particularly of the cock, are very bright and showy. The flesh is white, tender, and delicate, and on this account marketable; the eggs are small, and extremely delicate. The plumage is very beautiful—a clear, dark red, very bright, extending from the back to the extremities, while the breast is beautifully black. The upper convex side of the wing is equally red and black, and the whole of the tail-feathers white. The beak and legs are black; the eyes resemble jet beads, very full and brilliant; and the whole contour of the head gives a most ferocious expression.

THE GUELDERLAND.

The Guelderland fowl were originally imported into this country from the north of Holland, where they are supposed to have originated. They are very symmetrical in form, and graceful in their motions. They have one noticeable peculiarity, which consists in the absence of a comb in either sex. This is replaced by an indentation on the top of the head; and from the extreme end of this, at the back, a small spike of feathers rises. This adds greatly to the beauty of the fowl. The presence of the male is especially dignified, and the female is little inferior in carriage.

GUELDERLANDS.

The plumage is of a beautiful black, tinged with blue, of very rich appearance, and bearing a brilliant gloss. The legs are black, and, in some few instances, slightly feathered. Crosses with the Shanghae have heavily feathered legs. The wattles are of good size in the cock, while those of the hen are slightly less. The flesh is fine, of white color, and of excellent flavor. The eggs are large and delicate—the shell being thicker than in those of most other fowls—and are much prized for their good qualities. The hens are great layers, seldom inclining to sit. Their weight is from five pounds for the pullets, to seven pounds for the cocks.

The Guelderlands, in short, possess all the characteristics of a perfect breed; and in breeding them, this is demonstrated by the uniform aspect which is observable in their descendants. They are light and active birds, and are not surpassed, in point of beauty and utility, by any breed known in this country. The only objection, indeed, which has been raised against them is the tenderness of the chickens. With a degree of care, however, equal to their value, this difficulty can be surmounted, and the breed must be highly appreciated by all who have a taste for beauty, and who desire fine flesh and luscious eggs.

THE SPANGLED HAMBURGH.

The Spangled Hamburgh fowl are divided into two varieties, the distinctive characteristics being slight, and almost dependent upon color; these varieties are termed the Golden and Silver-Spangled. *The Golden Spangled* is one of no ordinary beauty; it is well and very neatly made, has a good body, and no very great offal. On the crest, immediately above the beak, are

HAMBURGH FOWLS.

two small, fleshy horns, resembling, to some extent, an abortive comb. Above this crest, and occupying the place of a comb, is a very large brown or yellow tuft, the feathers composing it darkening toward their extremities. Under the insertion of the lower mandible—or that portion of the neck corresponding to the chin in man—is a full, dark-colored tuft, somewhat resembling a beard. The wattles are very small; the comb, as in other high-crested fowls, is very diminutive; and the skin and flesh white. The hackles on the neck are of a brilliant orange, or golden yellow; and the general ground-color of the body is of the same hue, but somewhat darker. The thighs are of a dark-brown or blackish shade, and the legs and feet are of a bluish gray.

In the *Silver-spangled* variety, the only perceptible difference is, that the ground color is a silvery white. The extremity and a portion of the extreme margin of each feather are black, presenting, when in a state of rest, the appearance of regular semicircular marks, or spangles—and hence the name, "Spangled Hamburgh;" the varieties being termed *gold* or *silver*, according to the prevailing color being bright yellow, or silvery white.

The eggs are of moderate size, but abundant; chickens easily reared. In mere excellence of flesh and as layers, they are inferior to the Dorking or the Spanish. They weigh from four and a half to five and a half pounds for the male, and three and a half for the female. The former stands some twenty inches in height, and the latter about eighteen inches.

THE JAVA.

The Great Java fowl is seldom seen in this country in its purity. They are of a black or dark auburn color, with very large, thick legs, single comb and wattles. They are good layers, and their eggs are very large and well-flavored; their gait is slow and majestic. They are, in fact, amongst the most valuable fowls in the country, and are frequently described as Spanish fowls, than which nothing is more erroneous.

They are as distinctly an original breed as the pure-blooded Great Malay, and possess about the same qualities as to excellence, but fall rather short of them in beauty. Some, however, consider the pure Java superior to all other large fowls, so far as beauty is concerned. Their plumage is decidedly rich.

THE JERSEY-BLUE.

The color of this variety is light-blue, sometimes approaching to dun; the tail and wings rather shorter than those of the common fowl; its legs are of various colors, generally black, sometimes lightly feathered. Of superior specimens, the cocks weigh from seven to nine pounds, and the hens from six to eight.

They are evidently mongrels; and though once much esteemed, they have been quite neglected, so far as breeding from them is concerned, since the introduction of the purer breeds, as the Shanghaes and the Cochin-Chinas.

THE LARK-CRESTED FOWL.

This breed is sometimes confounded with the Polish fowl; but the shape of the crest, as well as the proportions of the bird, is different. This variety, of whatever color it may be, is of a peculiar taper-form, inclining forward, with a moderately depressed, backward-directed crest, and deficient in the neatness of the legs and feet so conspicuous in the Polands; the latter are of more upright carriage and of a more squarely-built frame. Perhaps a good distinction between the two varieties is, that the Lark-crested have an occipital crest, and the Poland more of a frontal one.

They are of various colors: pure snow white, brown with yellow hackles, and black. The white is, perhaps, more brilliant than is seen in any other domesticated gallinaceous bird, being much more dazzling than that of the White Guinea Fowl, or the White Pea Fowl. This white variety is in great esteem, having a remarkably neat and lively appearance when rambling about a homestead. They look very clean and attractive when dressed for market; an old bird, cleverly trussed, will be, apparently, as delicate and transparent in skin and flesh as an ordinary chicken. Their feathers are also more salable than those from darker colored fowls. They are but little, if, indeed, any, more tender than other kinds raised near the barn-door; they are in every way preferable to the White Dorkings.

In the cocks, a single, upright comb sometimes almost entirely takes the place of the crest; the hens, too, vary in this respect, some having not more than half a dozen feathers in their head-dress.

If they were not of average merit, as to their laying and sitting qualifications, they would not retain the favor they do with the thrifty house-wives by whom they are chiefly cultivated.

THE MALAY.

This majestic bird is found on the peninsula from which it derives its name, and, in the opinion of many, forms a connecting link between the wild and domesticated races of fowls.

MALAYS.

Something very like them is, indeed, still to be found in the East This native Indian bird—the *Gigantic Cock*, the *Kulm Cock* of Europeans—often stands considerably more than two feet from the crown of the head to the ground. The comb extends backward in a line with the eyes; it is thick, a little elevated, rounded upon the top, and has almost the appearance of having been cut off. The wattles of the under mandibles are comparatively small, and the throat is bare. Pale, golden-reddish hackles ornament the head, neck, and upper part of the back, and some of these spring before the bare part of the throat. The middle of the back and smaller wing-coverts are deep chestnut, the webs of the feathers disunited; pale reddish-yellow, long, drooping, hackles cover the rump and base of the tail, which last is very ample, and entirely of a glossy green, of which color are the wing-coverts; the secondaries and quills are pale reddish-yellow on their outer webs. All the under parts are deep glossy blackish-green, with high reflections; the deep chestnut of the base of the feathers appears occasionally, and gives a mottled and interrupted appearance to those parts.

The weight of the Malay, in general, exceeds that of the Cochin-China; the male weighing, when full-grown, from eleven to twelve, and even thirteen pounds, and the female from eight to ten pounds; height, from twenty-six to twenty-eight inches. They present no striking uniformity of plumage, being of all shades, from black to white; the more common color of the female is a light reddish-yellow, with sometimes a faint tinge of dunnish-blue, especially in the tail.

The cock is frequently of a yellowish-red color, with black intermingled in the breast, thighs, and tail. He has a small, but thick comb, generally inclined to one side; he should be snake-headed, and free from the slightest trace of top-knot; the wattles should be extremely small, even in an old bird; the legs are not feathered, as in the case of the Shanghaes, but, like them and the Cochin-Chinas, his tail is small, compared with his size. In the female, there is scarcely any show of comb or wattles. Their legs are long and stout; their flesh is very well flavored, when they have been properly fattened; and their eggs are so large and rich that two of them are equal to three of those of our ordinary fowls.

The Malay cock, in his perfection, is a remarkably courageous and strong bird. His beak is very thick, and he is a formidable antagonist when offended. His crow is loud, harsh, and prolonged, as in the case of the Cochin-China, but broken off abruptly at the termination; this is quite characteristic of the bird.

The chickens are at first very strong, with yellow legs, and are thickly covered with light brown down; but, by the time they are one-third grown, the increase of their bodies has so far outstripped that of their feathers, that they are half naked about the back and shoulders, and extremely susceptible of cold and wet. The great secret of rearing them is, to have them hatched very early indeed, so that they may have safely passed through this period of unclothed adolescence during the dry, sunny part of May and June, and reached nearly their full stature before the midsummer rains descend.

Malay hens are much used by some for hatching the eggs of turkeys—a task for which they are well adapted in every respect but one, which is, that they will follow their natural instinct in turning off their chickens at the usual time, instead of retaining charge of them as long as the mother turkey would have done. Goslings would suffer less from such untimely desertion.

The Pheasant Malay. This variety is highly valued by many, not on account of its intrinsic merits, which are considerable, but because it is believed to be a cross between the pheasant and the common fowl. This is, however, an erroneous opinion. Hybrids between the pheasant and the fowl are, for the most part, absolutely sterile; when they do breed, it is not with each other, but with the stock of one of their progenitors; and the offspring of these either fail or assimilate to one or the other original type. No half-bred family is perpetuated, no new breed created, by human or volucrine agency.

The Pheasant Malays are large, well-flavored, good sitters, good layers, good mothers, and, in many points, an ornamental and desirable stock. Some object to them as being a trifle too long in their make; but they have a healthy look of not being over-bred, which is a recommendation to those who rear for profit as well as pleasure. The eggs vary in size; some are very large in summer, smooth, but not polished, sometimes tinged with light buff, balloon-shaped, and without the zone of irregularity. The chickens, when first hatched, are all very much alike; yellow, with a black mark all down the back. The cock has a black tail, with black on the neck and wings.

THE PLYMOUTH ROCK.

This name has been given to a very good breed of fowls, produced by crossing a China cock with a hen, a cross between the Fawn-colored Dorking, the Great Malay, and the Wild Indian.

At a little over a year old, the cocks stand from thirty-two to thirty-five inches high, and weigh about ten pounds; and the pullets from six and a half to seven pounds each. The latter commence laying when five months old, and prove themselves very superior layers. Their eggs are of a medium size, rich, and reddish-yellow in color. Their plumage is rich and variegated; the cocks usually red or speckled, and the pullets darkish brown. They have very fine flesh, and are fit for the table at an early age. The legs are very large, and usually blue or green, but occasionally yellow or white, generally having five toes upon each foot. Some have their legs feathered, but this is not usual. They have large and single combs and wattles, large cheeks, rather short tails, and small wings in proportion to their bodies.

They are domestic, and not so destructive to gardens as smaller fowls. There is the same uniformity in size and general appearance, at the same age of the chickens, as in those of the pure bloods of primary races.

THE POLAND.

The Poland, or Polish fowl, is quite unknown in the country which would seem to have suggested the name, which originated from some fancied resemblance between its tufted crest and the square-spreading crown of the feathered caps worn by the Polish soldiers.

The breed of crested fowls is much esteemed by the curious, and is bred with great care. Those desirous of propagating any singular varieties, separate and confine the individuals, and do not suffer them to mingle with such as have the colors different. The varieties are more esteemed in proportion to the variety of the colors, or the contrast of the tuft with the rest of the plumage. Although the differences of plumage are thus preserved pretty constant, they seem to owe their origin to the same breed, and cannot be reproduced pure without careful superintendence. The cocks are much esteemed in Egypt, in consequence of the excellence of their flesh, and are so common that they are sold at a remarkably cheap rate. They are equally abundant at the Cape of Good Hope, where their legs are feathered.

The Polish are chiefly suited for keeping in a small way, and in a clean and grassy place. They are certainly not so fit for the farm-yard, as they become blinded and miserable with dirt. Care should be exercised to procure them genuine, since there is no breed of fowls more disfigured by mongrelism than this. They will, without any cross-breeding, occasionally produce white stock that are very pretty, and equally good for laying. If, however, an attempt is made to establish a separate breed of them, they become puny and weak. It is, therefore, better for those who wish for them to depend upon chance; every brood almost of the black produces one white chicken, as strong and lively as the rest.

POLAND FOWLS.

These fowls are excellent for the table, the flesh being white, tender, and juicy; but they are quite unsuitable for being reared in any numbers, or for general purposes, since they are so capricious in their growth, frequently remaining stationary in this respect for a whole month, getting no larger; and this, too, when they are about a quarter or half grown—the time of their life when they are most liable to disease. As aviary birds, they are unrivalled among fowls. Their plumage often requires a close inspection to appreciate its elaborate beauty; the confinement and fretting seem not uncongenial to their health; and their plumage improves in attractiveness with almost every month.

The great merit, however, of all the Polish fowls is, that for three or four years they continue to grow and gain in size, hardiness, and beauty—the male birds especially. This fact certainly points out a very wide deviation in constitution from those fowls which attain their full stature and perfect plumage in twelve or fifteen months The similarity of coloring in the two sexes—almost a specific distinction of Polish and perhaps Spanish fowls—also separates them from those breeds, like the Game, in which the cocks and hens are remarkably dissimilar. Their edible qualities are as superior, compared with other fowls, as their outward apparel surpasses in elegance. They have also the reputation of being everlasting layers, which further fits them for keeping in small enclosures; but, in this respect, individual exceptions are often encountered—as in the case of the Hamburghs—however truly the habit may be ascribed to the race.

There are four known varieties of the Polish fowl, one of which appears to be lost to this country.

The Black Polish. This variety is of a uniform black—both cock and hen—glossed with metallic green. The head is ornamented with a handsome crest of white feathers, springing from a fleshy protuberance, and fronted more or less deeply with black. The comb is merely two or three spikes, and the wattles are rather small. Both male and female are the same in color, except that the former has frequently narrow stripes of white in the waving feathers of the tail, a sign, it is said, of true breeding. The hens, also, have two or three feathers on each side of the tail, tinged in the tip with white. They do not lay quite so early in the spring as some varieties, especially after a hard winter; but they are exceedingly good layers, continuing a long time without wanting to sit, and laying rather large, very white, sub-ovate eggs. They will, however, sit at length, and prove of very diverse dispositions; some being excellent sitters and nurses, others heedless and spiteful.

The chickens, when first hatched, are dull black, with white breasts, and white down on the front of the head. They do not always grow and get out of harm's way so quickly as some other sorts, but are not particularly tender. In rearing a brood of these fowls, some of the hens may be observed with crests round and symmetrical as a ball, and others in which the feathers turn all ways, and fall loosely over the eyes; and in the cocks, also, some have the crest falling gracefully over the back of the head, and others have the feathers turning about and standing on end. These should be rejected, the chief beauty of the kind depending upon such little particulars. One hen of this variety laid just a hundred eggs, many of them on consecutive days, before wanting to incubate; and after rearing a brood successfully, she laid twenty-five eggs before moulting in autumn.

THE GOLDEN POLANDS. These are sometimes called Gold-spangled, as their plumage approaches to that of the Gold-Spangled Hamburghs; but many of the finest specimens have the feathers merely fringed with a darker color, and the cocks, more frequently than the hens, exhibit a spotted or spangled appearance. Many of them are disfigured by a muff or beard; as to which the question has been raised whether it is an original appendage to these birds or not. A distinct race, of which the muff is one permanent characteristic, is not at present known. This appendage, whenever introduced into the poultry-yard, is not easily got rid of; which has caused some to suspect either that the original Polish were beardless, or that there were two ancient races.

The Golden Polands, when well-bred, are exceedingly handsome; the cock has golden hackles, and gold and brown feathers on the back; breast and wings richly spotted with ochre and dark brown; tail darker; large golden and brown crest, falling back over the neck; but little comb and wattles. The hen is richly laced with dark-brown or black on an ochre ground; dark-spotted crest; legs light-blue, very cleanly made, and displaying a small web between the toes, almost as proportionately large as that in some of the waders.

They are good layers, and produce fair-sized eggs. Many of them make excellent mothers, although they cannot be induced to sit early in the season. The chickens are rather clumsy-looking little creatures, of a dingy-brown, with some dashes of ochre about the head, breast and wings. They are sometimes inclined to disease in the first week of their existence; but, if they pass this successfully, they become tolerably hardy, though liable to come to a pause when about half-grown. It may be noted as a peculiarity in the temper of this breed, that, if one is caught, or attacked by any animal, the rest, whether cocks or hens, will instantly make a furious attack upon the aggressor, and endeavor to effect the rescue of their companion.

THE SILVER POLANDS. These are similar to the preceding in shape and markings, except that white, black, and gray are exchanged for ochre or yellow, and various shades of brown. They are even more delicate in their constitution, more liable to remain stationary at a certain point of their adolescence, and, still more than the other varieties, require and will repay extra care and accommodation. Their top-knots are, perhaps, not so large, as a general thing; but they retain the same neat bluish legs and slightly-webbed feet. The hens are much more ornamental than the cocks; though the latter are sure to attract notice. They may, unquestionably, be ranked among the choicest of fowls, whether their beauty or their rarity is considered. They lay, in tolerable abundance, eggs of moderate size, French-white, much pointed at one end; and when they sit, acquit themselves respectably.

The newly-hatched chickens are very pretty; gray, with black eyes, light lead-colored legs, and a swelling of down on the crown of the head, indicative of the future top-knot, which is exactly the color of a powdered wig, and, indeed, gives the chicken the appearance of wearing one. There is no difficulty in rearing them for the first six weeks or two months; the critical time being the interval between that age and their reaching the fifth or sixth month. They acquire their peculiar distinctive features at a very early age, and are then the most elegant little miniature fowls which can possibly be imagined. The distinction of sex is not very manifest till they are nearly full-grown; the first observable indication being in the tail. That of the pullet is carried uprightly, as it ought to be; but in the cockerel, it remains depressed, awaiting the growth of the sickle-feathers. The top-knot of the cockerel inclines to hang more backward than that of the pullets. It is remarkable that the Golden Polish cock produces as true Silver chickens, and those stronger, with the Silver Polish hen, as the Silver Polish cock would bring.

The Silver Polands have all the habits of their golden companions, the main difference being the silvery ground instead of the golden. This variety will sometimes make its appearance even if merely its Golden kind is bred, precisely as the Black Polish now and then produce some pure White chickens that make very elegant birds.

THE BLACK-TOPPED WHITE. This variety does not at present exist among us; and some have even questioned whether it ever did. Buffon mentions them as if extant in France in his time. An attempt has been made to obtain them from the preceding, by acting on the imagination of the parents. The experiment failed, though similar schemes are said to have succeeded with animals; it proved, however, that it will not do to breed from the White Polish as a separate breed. Being Albinos, the chickens come very weakly, and few survive.

This breed is now recoverable, probably, only by importation from Asia.

THE SHANGHAE.

For all the purposes of a really good fowl—for beauty of model, good size, and laying qualities—the thorough-bred

Shanghae is among the best, and generally the most profitable of domestic birds. The cock, when full-grown, stands about twenty-eight inches high, if he is a good specimen; the female, about twenty-two or twenty-three inches. A large comb or heavy wattles are rarely seen on the hen at any age; but the comb of the male is high, deeply indented, and his wattles double and large. The comb and wattles are not, however, to be regarded as the chief characteristics of this variety, nor even its reddish-yellow feathered leg; but the abundant, soft, and downy covering of the thighs, hips, and region of the vent, together with the remarkably short tail, and large mound of feathers piled over the upper part of its root, giving rise to a considerable elevation on that part of the rump. It should be remarked, also, that the wings are quite short and small in proportion to the size of the fowl, and carried very high up the body, thus exposing the whole of the thighs, and a considerable portion of the side.

SHANGHAES.

These characteristics are not found, in the same degree, in any other fowl. The peculiar arrangement of feathers gives the Shanghae in appearance, what it has in reality—a greater depth of quarter, in proportion to the brisket, than any other fowl.

As to the legs, they are not very peculiar. The color is usually reddish-white, or flesh-color, or reddish-yellow, mostly covered down the outside, even to the end of the toes, with feathers. This last, however, is not always the case. The plumage of the thorough-bred is remarkably soft and silky, or rather downy; and is, in the opinion of many, equally as good for domestic purposes as that of the goose. The feathers are certainly quite as fine and soft, if not as abundant.

In laying qualities, the pure Shanghae equals, if it does not excel, any other fowl. The Black Poland, or the Bolton Gray, may, perhaps, lay a few more eggs in the course of a year, in consequence of not so frequently inclining to sit; but their eggs are not so rich and nutritious. A pullet of this breed laid one hundred and twenty eggs in one hundred and twenty-five days, then stopped six days, then laid sixteen eggs more, stopped four days, and again continued her laying. The eggs are generally of a pale yellow, or nankeen color, not remarkably large, compared with the size of the fowl, and generally blunt at the ends. The comb is commonly single, though, in some specimens, there is a slight tendency to rose.

The flesh of this fowl is tender, juicy, and unexceptionable in every respect. Taking into consideration the goodly size of the Shanghae—weighing, as the males do, at maturity, from ten to twelve pounds, and the females from seven and a half to eight and a half, and the males and females of six months, eight and six pounds respectively—the economical uses to which its soft, downy feathers may be applied, its productiveness, hardiness, and its quiet and docile temper, this variety must occupy, and deservedly so, a high rank among our domestic fowls; and the more it is known, the better will it be appreciated.

The White Shanghae. This variety is entirely white, with the legs usually feathered, and differ in no material

WHITE SHANGHAES.

respect from the red, yellow, and Dominique, except in color. The legs are yellowish, or reddish-yellow, and sometimes of flesh-color. Many prefer them to all others. The eggs are of a nankeen, or dull yellow color, and blunt at both ends.

It is claimed by the friends of this variety that they are larger and more quiet than other varieties, that their flesh is much superior, their eggs larger, and the hens more profitable. Being more quiet in their habits, and less inclined to ramble, the hens are invaluable as incubators and nurses; and the mildness of their disposition makes them excellent foster-mothers, as thev never injure the chickens belonging to other hens.

These fowls will rank among the largest coming from China, and are very thrifty in our climate. A cock of this variety attained a weight of eight pounds, at about the age of eight months, and the pullets of the same brood were proportionably large. They are broad on the back and breast, with a body well rounded up; the plumage white, with a downy softness —in the latter respect much like the feathering of the Bremen goose; the tail-feathers short and full; the head small, surmounted by a small, single, serrated comb; wattles long and wide, overlaying the cheek-piece, which is also large, and extends back on the neck; and the legs of a yellow hue, approaching a flesh-color, and feathered to the ends of the toes.

THE SILVER PHEASANT.

This variety of fowls is remarkable for great brilliancy of plumage and diversity of colors. On a white ground, which is usually termed silvery, there is an abundance of black spots. The feathers on the upper part of the head are much longer than the rest, and unite together in a tuft. They have a small, double comb, and their wattles are also comparatively small. A remarkable peculiarity of the cock is, that there is a spot of a blue color on the cheeks, and a range of feathers under the throat, which has the appearance of a collar.

The hen is a smaller bird, with plumage similar to that of the cock, and at a little distance seems to be covered with scales. On the head is a top-knot of very large size, which droops over it on every side. The Silver Pheasants are beautiful and showy birds, and chiefly valuable as ornamental appendages to the poultry-yard.

THE SPANISH.

This name is said to be a misnomer, as the breed in question was originally brought by the Spaniards from the West

SPANISH FOWLS.

Indies; and, although subsequently propagated in Spain, it has for some time been very difficult to procure good specimens from that country. From Spain, they were taken in considerable numbers into Holland, where they have been carefully bred, for many years; and it is from that quarter that our best fowls of this variety come.

The Spanish is a noble race of fowls, possessing many merits; of spirited and animated appearance; of considerable size; excellent for the table, both in whiteness of flesh and skin, and also in flavor; and laying exceedingly large eggs in considerable numbers. Among birds of its own breed it is not deficient in courage; though it yields, without showing much fight, to those which have a dash of game blood in their veins. It is a general favorite in all large cities, for the additional advantage that no soil of smoke or dirt is apparent on its plumage.

The thorough-bred birds should be entirely black, as far as feathers are concerned; and when in high condition, display a greenish, metallic lustre. The combs of both cock and hen are exceedingly large, of a vivid and most brilliant scarlet; that of the hen droops over upon one side. Their most singular feature is a large, white patch, or ear-hole, on the cheek—in some specimens extending over a great part of the face—of a fleshy substance, similar to the wattle; it is small in the female, but large and very conspicuous in the male. This marked contrast of black, bright red, and white, makes the breed of the Spanish cock as handsome as that of any variety which we have; in the genuine breed, the whole form is equally good.

Spanish hens are celebrated as good layers, and produce very large, quite white eggs, of a peculiar shape, being very thick at both ends, and yet tapering off a little at each. They are, by no means, good mothers of families, even when they do sit—which they will not often condescend to do—proving very careless, and frequently trampling half their brood under foot. The inconveniences of this habit are, however, easily obviated by causing the eggs to be hatched by some more motherly hen.

This variety of fowl has frequently been known to lose nearly all the feathers in its body, besides the usual quantity on the neck, wings, and tail; and, if they moult late, and the weather is severe, they feel it much. This must often happen in the case of an "everlasting layer;" for if the system of a bird is exhausted by the unremitting production of eggs, it cannot contain within itself the material for supplying the growth of feathers. They have not, even yet, become acclimated in this country, since continued frost at any time is productive of much injury to their combs; frequently causing mortification in the end, which at times terminates in death. A warm poultry-house, high feeding, and care that they do not remain too long exposed to severe weather, are the best means of preventing this disfigurement. Some birds are occasionally produced, handsomely streaked with red on the hackle and back. This is no proof of bad breeding, if other points are right.

The chickens are large, as would be expected from such eggs, entirely shining black, except a pinafore of white on the breast—in which respect they are precisely like the Black Polish chickens—and a slight sprinkling under the chin, with sometimes also a little white round the back and eyes; their legs and feet are black. Many of them do not get perfectly feathered till they are three-fourths grown; and, therefore, to have this variety come to perfection in a country where the summers are much shorter than in their native climate, they must be hatched early in spring, so that they may be well covered with plumage before the cold rains of autumn. There is, however, a great lack of uniformity in the time when they get their plumage; the pullets are always earlier and better feathered than the cockerels—the latter being generally half naked for a considerable time after being hatched, though some feather tolerably well at an early age.

The *Black* is not the only valuable race of Spanish fowl; there is, also, the *Gray*, or *Speckled*, of a slaty gray color, with white legs. Their growth is so rapid, and their size, eventually, so large, that they are remarkably slow in obtaining their feathers. Although well covered with down when first hatched, they look almost naked when half-grown, and should, therefore, be hatched as early in spring as possible. The cross between the Pheasant-Malay and the Spanish produces a particularly handsome fowl.

As early pullets, for laying purposes in the autumn and winter after they are hatched, no fowls can surpass the Spanish. They are believed, also, to be more precocious in their constitution; and consequently to lay at an earlier age than the pullets of other breeds.

THE NATURAL HISTORY OF DOMESTIC FOWLS.

Fowls are classed by modern naturalists as follows:

DIVISION. *Vertebrata*—possessing a back bone.

CLASS. *Aves*—birds.

ORDER. *Rasores*—scrapers.

FAMILY. *Phasianidæ*—Pheasants.

GENUS. *Gallus*—the cock.

Birds, as well as quadrupeds, may be divided into two great classes, according to their food: the Carnivorous and the Graminivorous. Fowls belong, strictly speaking, to the latter.

In the structure of the *digestive organs*, birds exhibit a great uniformity. The œsophagus, which is often very muscular, is dilated into a large sac—called the *crop*—at its entrance into the breast; this is abundantly supplied with glands, and serves as a species of first stomach, in which the food receives a certain amount of preparation before being submitted to the action of the proper digestive organs. A little below the crop, the narrow œsophagus is again slightly dilated, forming what is called the *ventriculus succenturiatus*, the walls of which are very thick, and contain a great number of glands, which secrete the gastric juice. Below this, the intestinal canal is enlarged into a third stomach, the *gizzard*, in which the process of digestion is carried still farther. In the graminivorous birds, the walls of this cavity are very thick and muscular, and clothed internally with a strong, horny *epithelium*, serving for the trituration of the food. The intestine is rather short, but usually exhibits several convolutions; the large intestine is always furnished with two *corea*. It opens by a semicircular orifice into the *cloaca*, which also receives the orifices of the urinary and generative organs. The liver is of large size, and usually furnished with a gall-bladder. The pancreas is lodged in a kind of loop, formed by the small intestine immediately after quitting the gizzard. There are also large salivary glands in the neighborhood of the mouth, which pour their secretion into that cavity.

The *organs of circulation and respiration* in birds are adapted to their peculiar mode of life; but are not separated from the abdominal cavity by a diaphragm, as in the mammalia. The heart consists of four cavities distinctly separated—two auricles and two ventricles—so that the venous and arterial blood can never mix in that organ; and the whole of the blood returned from the different parts of the body passes through the lungs before being again driven into the systemic arteries. The blood is received from the veins of the body in the right auricle, from which it passes through a tabular opening into the right ventricle, and is thence driven into the lungs. From these organs it returns through the pulmonary veins into the left auricle, and passes thence into the ventricles of the same side, by the contraction of which it is driven into the aorta. This soon divides into two branches, which, by their subdivision, give rise to the arteries of the body.

The jaws, or mandibles, are sheathed in a horny case, usually of a conical form, on the sides of which are the nostrils. In most birds, the sides of this sheath or bill are smooth and sharp; but in some they are denticulated along the margins. The two anterior members of the body are extended into wings. The beak is used instead of hands; and such is the flexibility of the vertebral column, that the bird is able to touch with its beak every part of its body. This curious and important result is obtained chiefly by the lengthened vertebræ of the neck, which, in the swan, consists of twenty-three bones, and in the domestic cock, thirteen. The vertebræ of the back are seven to eleven; the ribs never exceed ten on each side.

The clothing of the skin consists of *feathers*, which in their nature and development resemble hair, but are of a more complicated structure. A perfect feather consists of the *shaft*, a central stem, which is tubular at the base, where it is inserted into the skin, and the *barbs*, or fibres, which form the *webs* on each side of the shaft. The two principal modifications of feathers are *quills* and *plumes;* the former confined to the wings and tail, the latter constituting the general clothing of the body. Besides the common feathers, the skin of many birds is covered with a thick coating of down, which consists of a multitude of small feathers of peculiar construction; each of these down feathers is composed of a very small, soft tube imbedded in the skin, from the interior of which there rises a small tuft of soft filaments, without any central shaft. These filaments are very slender, and bear on each side a series of still more delicate filaments, which may be regarded as analogous to the barbules of the ordinary feathers. This downy coat fulfils the same office as the soft, woolly fur of many quadrupeds; the ordinary feathers being analogous to the long, smooth hair by which the fur of these animals is concealed. The skin also bears many hair-like appendages, which are usually scattered sparingly over its surface; they rise from a bulb which is imbedded in the skin, and usually indicate their relation to the ordinary feathers by the presence of a few minute barbs toward the apex.

Once or twice in the course of a year the whole plumage of the bird is renewed, the casting of the old feathers being called *moulting.* The base of the quills is covered by a series of large feathers, called the *wing coverts;* and the feathers of the tail are furnished with numerous muscles, by which they can be spread out and folded up like a fan. In the aquatic birds—like the goose, the duck, and the swan—the feathers are constantly lubricated by an oily secretion, which completely excludes the water.

In their reproduction, birds are strictly oviparous. The *eggs* are always enclosed in a hard shell, consisting of calcareous matter, and birds almost invariably devote their whole attention, during the breeding season, to the hatching of their eggs and the development of their offspring; sitting constantly upon the eggs to communicate to them the degree of warmth necessary for the evolution of the embryo, and attending to the wants of their newly-hatched young, until the latter are in a condition to shift for themselves.

In the structure and development of the egg there is a great uniformity; but there is a remarkable difference in the condition of the young bird at the moment of hatching. In the class under consideration, the young are able to run about from the moment of their breaking the egg-shell; and the only care of the parents is devoted to protecting their offspring from danger, and leading them into those places where they are likely to meet with food.

The *longevity* of birds is various, and, unlike the case of men and quadrupeds, seems to bear little proportion to the age at which they acquire maturity. A few months, or even a few weeks, suffice to bring them to their perfection of stature, instincts, and powers. Domestic fowls live to the age of twenty years; geese, fifty; while swans exceed a century.

The order *Rasores* includes the numerous species of *gallinaceous birds*, and the term is applied to them from their habit of scratching in the ground in search of food. They are generally marked by a small head, stout legs, plumage fine, the males usually adorned with magnificent colors, and the tails often developed in a manner to render the appearance extremely elegant. The wings are usually short and weak, and the flight of the birds is neither powerful nor prolonged. The *corla* of this order are larger than in any other birds.

The species are found in almost all parts of the world, from the tropics to the frozen regions of the north; but the finest and most typical kinds are inhabitants of the temperate and warmer parts of Asia. They feed principally on seeds, fruit, and herbage, but also, to a considerable extent, on insects, worms, and other small animals. Their general habitation is on the ground, where they run with great celerity, but many of them roost on trees. They are mostly polygamous in their habits, the males being usually surrounded by a considerable troop of females; and to these, with a few exceptions, the whole business of incubation is generally left. The nest is always placed on the ground in some sheltered situation, and very little art is exhibited in its construction; indeed, an elaborate nest is the less necessary, as the young are able to run about and feed almost as soon as they have left the egg; and at night or on the approach of danger, they collect beneath the wings of their mother. Most of these species are esteemed for the table, and many of them are among the most celebrated of game birds.

The *pheasant family*, of this order, includes the most beautiful of the rasorial birds; indeed, some of them may, perhaps, be justly regarded as pre-eminent in this respect over all the rest of their class. In these, the bill is of moderate size and compressed, with the upper mandible arched to the tip, where it overhangs the lower one; the *tarsi* are of moderate length and thickness, usually armed with one or two spurs; the toes are moderate, and the hinder one short and elevated; the wings are rather short and rounded, and the tail more or less elongated and broad, but frequently wedge-shaped and pointed. The head is rarely feathered all over; the naked skin is sometimes confined to a space about the eye, but generally occupies a greater portion of the surface, occasionally covering the whole head, and even a part of the neck, and frequently forming combs and wattles of very remarkable forms. In some species, the crown is furnished with a crest of feathers.

The birds of this family are, for the most part, indigenous to the Asiatic continent and islands, from which, however, several species have been introduced into other parts of the globe. The Guinea Fowl of Africa, and the Turkeys of America, are almost the only instances of wild Phasianidous birds out of Asia. Some species, such as the Domestic Fowl, the Peacock, the Turkey, and the Guinea Fowl, have been reduced to a state of complete domestication, and are distributed pretty generally over the world.

THE GUINEA FOWL.

This bird belongs to the same division, class, order, and family as the Domestic Fowl; but is assigned by naturalists to the genus Numida, or Numidian. It is indigenous to the tropical parts of Africa, and in a wild state, Guinea Fowls live in flocks, in woods, preferring marshy places, and feed on insects, worms, and seeds; they roost on trees; the nest is made on the ground, and usually contains as many as twenty eggs. They have been propagated in the Island of Jamaica to such an extent as to have become wild, and are shot like other game. They do much damage to the crops, and are therefore destroyed by various means; one of which is, to get them tipsy by strewing corn steeped in rum, and mixed with the intoxicating juice of the cassava, upon the ground; the birds devour this, and are soon found in a helpless state of inebriety.

The Guinea Fowl, to a certain degree, unites the characteristics of the pheasant and the turkey; having the delicate shape of the one, and the bare head of the other. There are several varieties: the White, the Spotted, the Madagascar, and the Crested. The latter is not so large as the common species; the head and neck are bare, of a dull blue, shaded with red, and, instead of the casque, it has an ample crest of hair-like, disunited feathers, of a bluish black, reaching as far forward as the nostrils, but, in general, turned backward. The whole plumage, except the quills, is of a bluish black, covered with small grayish spots, sometimes four, sometimes six on each feather.

THE GUINEA FOWL.

This fowl is not a great favorite among many keepers of poultry, being so unfortunate as to have gained a much worse reputation than it really deserves, from having been occasionally guilty of a few trifling faults. It is, however, useful, ornamental, and interesting during its life; and, when dead, a desirable addition to the table, at a time when all other poultry is scarce.

The best way to commence keeping Guineas is to procure a sitting of eggs which can be depended upon for freshness, and, if possible, from a place where but a single pair is kept. A Bantam hen is the best mother; she is lighter, and less likely to injure them by treading on them than a full-sized fowl. She will cover nine eggs, and incubation will last a month. The young are excessively pretty. When first hatched, they are so strong and active as to appear not to require the attention which is really necessary to rear them. Almost as soon as they are dry from the moisture of the egg, they will peck each other's toes, as if supposing them to be worms, scramble with each other for a crumb of bread, and domineer over any little Bantam or chicken that may chance to have been hatched at the same time with themselves. No one, ignorant of the fact, would guess, from their appearance, to what species of bird they belonged; their orange-red bills and legs, and the dark, zebra-like stripes with which they are regularly marked from head to tail, bear no traces of the speckled plumage of their parents.

Of all known birds, the Guinea fowl is, perhaps, the most prolific of eggs. Week after week, and month after month, there are very few intermissions, if any, of the daily deposit. Even the process of moulting is sometimes insufficient to draw off the nutriment which it takes to make feathers instead of eggs; and the poor thing will sometimes go about half-naked in the chilly autumn months, unable to refrain from its diurnal visit to the nest, and consequently unable to furnish itself with a new outer garment. The body of the Guinea hen may be regarded, in fact, as a most admirable machine for producing eggs out of insects, grain, and vegetables, garbage, or whatever material an omnivorous creature can appropriate.

Its normal plumage is singularly beautiful, being spangled over with an infinity of white spots on a black ground, shaded with gray and brown. The spots vary from the size of a pea to extreme minuteness. The black and white occasionally change places, causing the bird to appear covered with a net of lace.

The white variety is not uncommon, and is said to be equally hardy and profitable with the usual kind; but the peculiar beauty of the original plumage is, certainly, all exchanged for a dress of not the purest white. It is doubtful how long either this or the former variety would remain permanent; though, probably, but for a few generations. Pied birds blotched with patches of white, are frequent, but are not comparable, in point of beauty, with those of the original wild color.

THE PEA FOWL.

This bird is assigned to the genus *paro*, or peacock—the division, class, or sex, and family, being the same as the preceding. The male of this species is noted for its long, lustrous tail, which it occasionally spreads, glittering with hundreds of jewel-like eye-spots, producing an unrivalled effect of grace and beauty. The form of the bird is also exceedingly elegant, and the general plumage of the body exhibits rich metallic tints; that of the neck, particularly, being of a fine deep blue, tinged with golden green. The female, however, is of a much more sober hue, her whole plumage being usually of a brownish color. The voice of the peacock is by no means suitable to the beauty of its external appearance, consisting of a harsh, disagreeable cry, not unlike the word *paon*, which is the French name of the bird.

Although naturalized as a domestic bird in Europe and America, the pea fowl is a native of India, where it is still found abundantly in a wild state; and the wild specimens are said to be more brilliant than those bred in captivity. The date of its introduction into England is not known; but the first peacocks appear to have been brought into Europe by Alexander the Great, although these birds were among the articles imported into Judea by the fleets of Solomon. They reached Rome toward the end of the republic, and their costliness soon caused them to be regarded as one of the greatest luxuries of the table, though the moderns find them dry and leathery. This, perhaps, as much as the desire of ostentation, may have induced the extravagance of Vitellius and Heliogabulus, who introduced dishes composed only of the brains and tongues of peacocks at their feasts. In Europe, during the middle ages, the peacock was still a favorite article in the bill of fare of grand entertainments, at which it was served with

THE PEA FOWL.

the greatest pomp and magnificence. And during the period of chivalry, it was usual for knights to make vows of enterprize on these occasions, "before the peacock and the ladies." At present, however, the bird is kept entirely on account of the beauty of its appearance.

In a state of nature, pea fowl frequent jungles and wooded localities, feeding upon grain, fruits, and insects. They are polygamous, and the females make their nests upon the ground among bushes; the nest is composed of grass, and the number of eggs laid is said to be five or six. They roost in high trees, and, even in captivity, their inclination to get into an elevated position frequently manifests itself; and they may often be seen perched upon high walls, or upon the ridges of buildings.

The latter characteristic is, indirectly, one reason why many are disinclined to keep pea fowl in a domestic state. Their decided determination so to roost prevents such a control being exercised over them as would restrain them from mischief, until an eye could be kept on their movements; and, consequently, they commit many depredations upon gardens, stealing off to their work of plunder at the first dawn, or at the most unexpected moments. Their cunning indeed is such that, if frequently driven away from the garden at any particular hour of the day or evening, they will never be found there, after a certain time, at that special hour, but will invariably make their inroads at day-break. Many have tried, as a last resort, to eject them with every mark of scorn and insult, such as harsh words, the cracking of whips, and the throwing of harmless brooms; but they remain incorrigible marauders, indifferent to this disrespectful usage, and careless of severe rebuke.

A mansion, therefore, where the fruit and vegetable garden is at a distance, is almost the only place where they can be kept without daily vexation. The injury they do to flowers is comparatively trifling; though, like the Guinea-fowl, they are great eaters of buds, cutting them out cleanly from the *axillae* of leaves. They must likewise have a dusting-hole, which is large and unsightly; but this can be provided for them in some nook out of the way; and by feeding and encouragement, they will soon be taught to dispose themselves into a pleasing spectacle, at whatever point of view may be deemed desirable. No one with a very limited range should attempt to keep them at all, unless confined in an aviary. Where they can be kept at large, they should be collected in considerable numbers, that their dazzling effects may be as impressive as possible.

A wanton destructiveness toward the young of other poultry is also charged upon them. Relative to this, however, statements differ; some contending that such instances of cruelty constitute the exceptions, and not the rule.

The hen does not lay till her third summer; but she then seems to have an instinctive fear of her mate, manifested by the secrecy with which she selects the place for her nest; nor, if the eggs are disturbed, will she go there again. She lays from four or five to seven. If these are taken, she will frequently lay a second time during the summer; and the plan is recommended to those who are anxious to increase their stock. She sits from twenty-seven to twenty-nine days. A common hen will hatch and rear the young; but the same objection lies against her performing that office, except in very fine, long summers, for the pea fowl as for turkeys—that the young require to be brooded longer than the hen is conveniently able to do. A turkey will prove a much better foster-mother in every respect. The peahen should, of course, be permitted to take charge of one set of eggs. Even without such assistance, she will be tolerably successful.

The same wise provision of Nature noticed in the case of the Guinea fowl is evinced in a still greater degree in the little pea chickens. Their native jungle—tall, dense, sometimes impervious, swarming with reptile, quadruped, and even insect, enemies—would be a most dangerous habitation for a little tender thing that could merely run and squall. Accordingly, they escape from the egg with their quill-feathers very highly developed. In three days, they will fly up, and perch upon any thing three feet high; in a fortnight, they will roost on trees, or the tops of sheds; and in a month or six weeks, they will reach the ridge of a barn, if there are any intermediate low stables or other buildings to help them to mount from one to the other.

There are two varieties of the common pea fowl: the *pied* and the *white*. The first has irregular patches of white about it, like the pied Guinea fowl, and the remainder of the plumage resembling the original sorts; the white have the ocellated spots on the tail faintly visible in certain lights. These last are tender, and much prized by those who prefer variety to real beauty. They are occasionally produced by birds of the common kind, in cases where no intercourse with other white birds can have taken place. In one instance, in the same brood, whose parents were both of the usual colors, there were two of the common sort, and one white cock, and one white hen.

THE TURKEY.

THE WILD TURKEY. The turkey belongs to the genus *meleagris*, and, though now known as a domestic fowl in most civilized countries, was confined to America until after the discovery of that country by Columbus. It was probably introduced into Europe by the Spaniards about the year 1530. It was found in the forests of North America, when the country was first settled, from the Isthmus of Darien to Canada, being then abundant even in New England; at present, a few are found in the mountains of Massachusetts, New York, and New Jersey; in the Western and the South-western States they are still numerous, though constantly diminishing before the extending and increasing settlements.

THE WILD TURKEY.

The wild male bird measures about three feet and a half, or

nearly four feet, in length, and almost six in expanse of the wings, and weighs from fifteen to forty pounds. The skin of the head is of a bluish color, as is also the upper part of the neck, and is marked with numerous reddish, warty elevations, with a few black hairs scattered here and there. On the under part of the neck, the skin hangs down loosely, and forms a sort of wattle; and from the point where the bill commences, and the forehead terminates, arises a fleshy protuberance, with a small tuft of hair at the extremity, which becomes greatly elongated when the bird is excited; and at the lower part of the neck is a tuft of black hair, eight or nine inches in length.

The feathers are, at the base, of a bright dusky tinge, succeeded by a brilliant metallic band, which changes, according to the point whence the light falls upon it, to bronze, copper, violet, or purple; and the tip is formed by a narrow, black, velvety band. This last marking is absent from the neck and breast. The color of the tail is brown, mottled with black, and crossed with numerous lines of the latter color; near the tip is a broad, black band, then a short mottled portion, and then a broad band of dingy yellow. The wings are white, banded closely with black, and shaded with brownish yellow, which deepens in tint toward the back. The head is very small, in proportion to the size of the body; the legs and feet are strongly made, and furnished with blunt spurs, about an inch long, and of a dusky reddish color; the bill is reddish, and brown-colored at the tip.

The female is less in size; her legs are destitute of spurs; her neck and head are less naked, being furnished with short, dirty, gray feathers; the feathers on the back of the neck have brownish tips, producing on that part a brown, longitudinal band. She also, frequently, but not invariably, wants the tuft of feathers on the breast. Her prevailing color is a dusky gray, each feather having a metallic band, less brilliant than that of the cock, then a blackish band, and a grayish fringe. Her whole color is, as usual among birds, duller than that of the cock; the wing-feathers display the white, and have no bands; the tail is similarly colored to that of the cock. When young, the sexes are so much alike that it is not easy to discern the difference between them; and the cock acquires his beauty only by degrees, his plumage not arriving at perfection until the fourth or fifth year.

The habits of these birds in their native wilds are exceedingly curious. The males, called *Gobblers*, associate in parties of from ten to a hundred, and seek their food apart from the females, which either go about singly with their young, at that time about two-thirds grown, or form troops with other females and their families, sometimes to the number of seventy or eighty. These all avoid the old males, who attack and destroy the young, whenever they can, by reiterated blows upon the skull. But all parties travel in the same direction, and on foot, unless the dog or the hunter or a river on their line of march compels them to take wing. When about to cross a river, they select the highest eminences, that their flight may be more sure, and in such positions they sometimes stay for a day or more, as if in consultation. The males upon such occasions gobble obstreperously, strutting with extraordinary importance, as if to animate their companions; and the females and the young assume much of the same pompous manner, and spread their tails as they move silently around. Having mounted, at length, to the tops of the highest trees, the assembled multitude, at the signal note of their leader, wing their way to the opposite shore. The old and fat birds, contrary to what might be expected, cross without difficulty, even when the river is a mile in width; but the wings of the young and the meagre, and, of course, those of the weak, frequently fail them before they have completed their passage, when they drop in, and are forced to swim for their lives, which they do cleverly enough, spreading their tails for a support, closing their wings, stretching out their necks, and striking out quickly and strongly with their feet. All, however, do not succeed in such attempts, and the weaker often perish.

The wild turkeys feed on maize, all sorts of berries, fruits, grasses, and beetles; tadpoles, young frogs, and lizards, are occasionally found in their crops. The pecan nut is a favorite food, and so is the acorn, on which last they fatten rapidly. About the beginning of October, while the mast still hangs on the trees, they gather together in flocks, directing their course to the rich bottom-lands, and are then seen in great numbers on the Ohio and Mississippi. This is the *turkey-month* of the Indians. When they have arrived at the land of abundance, they disperse in small, promiscuous flocks of both sexes and every age, devouring all the mast as they advance. Thus they pass the autumn and winter, becoming comparatively familiar after their journeys, when they venture near plantations and farm-houses. They have even been known, on such occasions, to enter stables and corn-cribs in quest of food. Numbers are killed in the winter, and preserved in a frozen state for distant markets.

The beginning of March is the pairing season, for a short time previous to which the females separate from their mates, and shun them, though the latter pertinaciously follow them, gobbling loudly. The sexes roost apart, but at no great distance, so that when the female utters a call, every male within hearing responds, rolling note after note in the most rapid succession; not as when spreading the tail and strutting near the hen, but in a voice resembling that of the tame turkey when he hears any unusual or frequently-repeated noise.

Where the turkeys are numerous, the woods, from one end to the other, sometimes for hundreds of miles, resound with this remarkable voice of their wooing, uttered responsively from their roosting-places. This is continued for about an hour; and, on the rising of the sun, they silently descend from their perches, and the males begin to strut for the purpose of winning the admiration of their mates.

If the call of a female be given from the ground, the males in the vicinity fly toward the individual, and, whether they perceive her or not, erect and spread their tails, throw the head backward, and distend the comb and wattles, shout pompously, and rustle their wings and body-feathers, at the same moment ejecting a puff of air from the lungs. While thus occupied, they occasionally halt to look out for the female, and then resume their strutting and puffing, moving with as much rapidity as the nature of their gait will admit. During this ceremonious approach, the males often encounter each other, and desperate battles ensue, when the conflict is only terminated by the flight or death of the vanquished. The usual fruits of such victories are reaped by the conqueror, who is followed by one or more females, that roost near him, if not upon the same tree, until they begin to lay, when their habits are altered, with the view of saving their eggs, which the male breaks, if he can get at them. These are usually from nine to fifteen in number, sometimes twenty, whitish and spotted with brown, like those of the domestic bird. The nest consists of a few dried leaves placed on the ground, sometimes on a dry ridge, sometimes on the fallen top of a dead leafy tree, under a thicket of sumach or briers, or by the side of a log. Whenever the female leaves the nest, she covers it with leaves, so as to screen it from observation. She is a very close sitter, and when she has chosen a spot will seldom leave it, on account of its being discovered by a human intruder. Should she find one of her eggs, however, sucked by a snake, or other enemy, she abandons the nest forever. When the eggs are near hatching, she will not forsake her nest while life remains.

The females are particularly attentive to their young, which are very sensitive to the effects of damp; and consequently wild turkeys are always scarce after a rainy season. The flesh of the wild turkey is much superior to that of the domestic bird; yet the flesh of such of the latter as have been suffered to roam at large in the woods and in the plains is, in no respect, improved by this partially wild mode of life.

THE DOMESTIC TURKEY.

The origin of the popular name, turkey, appears to be the confusion at first unaccountably subsisting relative to the identity of the bird with the Guinea fowl, which was still scarce at the time of the introduction of the turkey. Some, however, say that the name arose from the proud and *Turkish* strut of the cock. There is a question whether the domestic turkey is actually a second and distinct species, or merely a variety of the wild bird, owing its diversity of aspect to circumstances dependent on locality, and consequent change of habit, combined with difference of climate and other important causes, which are known in the case of other animals to produce such remarkable effects.

THE DOMESTIC TURKEY.

The *varieties* of the domesticated turkey are not very distinct; and as to their relative value, it is, perhaps, difficult to give any decisive opinion. Some suppose that the *white* turkey is the most robust, and most easily fattened. Experience has, however, shown to the contrary. The pure white are very elegant creatures; and though very tender to rear, are not so much so as the white pea fowl. Most birds, wild as well as tame, occasionally produce perfectly white individuals, of more delicate constitution than their parents. The selection and pairing of such have probably been the means of establishing and keeping up this breed. With all care, they will now and then produce speckled birds and so show a tendency to return to the normal plumage. It is remarkable that in specimens which are, in other respects, snow-white, the tuft on the breast remains coal-black, appearing, in the hens, like a tail of ermine, and so showing us a great ornament. The head and caruncles on the neck of the male are, when excited, of the same blue and scarlet hues. The bird is truly beautiful, with its snowy and trembling flakes of plumage thus relieved with small portions of black, blue, and scarlet. They have one merit—they dress most temptingly white for market; but they are unsuited for mirey, smokey, or clayey situations, and show and thrive best where they have a range of clean, short pasture, on a light or chalky subsoil.

The *bronze* and *copper-colored* varieties are generally undersized, and are among the most difficult of all to rear; but their flesh is, certainly, very delicate, and, perhaps, more so than that of other kinds—a circumstance, however, that may partly result from their far greater delicacy of constitution, and the consequent extra trouble devoted to their management.

The *brown* and *ashy-gray* are not particularly remarkable; but the *black* are decidedly superior, in every respect, not only as regards greater hardiness, and a consequent greater facility of rearing, but as acquiring flesh more readily, and that, too, of the very best and primest quality. Those of this color appear also to be far less removed than the others from the original wild stock. Fortunately, the black seems to be the favorite color of Nature; and black turkeys are produced far more abundantly than those of any other hue.

The turkey is a most profitable bird, since it can almost wholly provide for itself about the roads; snails, slugs, and worms are among the number of its dainties, and the nearest stream serves to slake its thirst. To the farmer, however, it is often a perfect nuisance, from its love of grain; and should, therefore, be kept in the yard until all corn is too strong in the root to present any temptation.

Notwithstanding the separation which, with the exception of certain seasons, subsists between the cock and hen turkey in a wild state, they have been taught to feed and live amiably together in a state of domesticity. The former, however, retains sufficient of his hereditary propensities to give an occasional sly blow to a froward chicken, but that very seldom of a serious or malicious character.

One reason why the turkeys seen in poultry-yards do not vie in splendor of plumage with their untamed brethren is, that they are not allowed to live long enough. For the same cause, the thorough development of their temper and disposition is seldom witnessed. It does not attain its full growth till its fifth or sixth year, yet it is killed at latest in the second, to the evident deterioration of the stock. If some of the best breeds were retained to their really adult state, and well fed meanwhile, they would quite recompense their keeper by their beauty in full plumage, their glancing hues of gilded green and purple, their lovely shades of bronze, brown, and black, and the pearly lustre that radiates from their polished feathers.

THE DUCK.

This bird is of the order of *natatores*, or swimmers; family, *anatidæ*, of the duck kind; genus, *anas*, or duck. The most striking character of the swimming bird is derived from the structure of the *feet*, which are always palmate—that is, furnished with webs between the toes. There are always three toes directed forward, and these are usually united by a membrane to their extremities; but, in some cases, the membrane is deeply cleft, and the toes are occasionally quite free, and furnished with a distinct web on each side. The fourth toe is generally but little developed, and often entirely wanting; when present, it is usually directed backward, and the membrane is sometimes continued to it along the side of the feet. These webbed feet are the principal agents by which the birds propel themselves through the water, upon the surface of which most of them pass a great portion of their time. The feet are generally placed very far back, a position which is exceedingly favorable to their action in swimming, but which renders their progression on the land somewhat awkward.

THE EIDER DUCK.

The *body* is generally stout and heavy, and covered with a very thick, close, downy plumage, which the bird keeps constantly anointed with the greasy secretions of the caudal gland, so that it is completely water-proof. The *wings* exhibit a great variety in their development; in some species being merely rudimentary, destitute of quills, and covered with a scaly skin—in others, being of vast size and power, and the birds passing a great part of their lives in the air. The form of the *bill* is also very remarkable; in some, broad and flat; in others, deep and compressed; and in others, long and slender.

Most of these birds live in societies, which are often exceedingly numerous, inhabiting high northern and southern latitudes.

The distinguishing characteristic of the family of the *anatidæ* is the *bill*, which is usually of a flattened form, covered with a soft skin, and furnished at the edges with a series of *lamellæ*, which serve to sift or strain the mud in which they generally seek their food. The feet are furnished with four toes, three of which are directed forward, and united by a web; the fourth is directed backward, usually of small size, and quite free. They are admirable swimmers, and live and move on the water with the utmost security, ease, and grace. Such is their adaptation to this element that the young, immediately after being hatched, will run to it, and fearlessly launch themselves upon its bosom, rowing themselves along with their webbed feet, without a single lesson, and yet as dexterously as the most experienced boatman. They are generally inhabitants of the fresh waters, and for the most part, prefer ponds and shallow lakes, in which they can investigate the bottom with their peculiar bills, without actually diving beneath the surface; yet at some seasons they are found along the borders of the sea. Their food generally consists of worms, mollusca, and aquatic insects, which they separate from the mud by the agency of the lamellæ at the margin of the bill; but most of them also feed upon seeds, fruits, and other vegetable substances.

THE WILD DUCK.

This bird, known also by the name of *mallard*, is the original of all the domestic varieties. It is twenty-four inches long, and marked with green, chestnut and white. Wild ducks are gregarious in their habits, and generally migrate in large flocks. The males are larger than the females, and the letter are also usually of a more uniform and sober tint.

It is an inhabitant of all the countries of Europe, especially toward the north, and is also abundant in North America,

WILD DUCK.

where it is migratory, passing to the North in Spring, and returning to the South in autumn. It frequents the lakes of the interior, as well as the sea-coasts. It is plentiful in Great Britain at all seasons, merely quitting the more exposed situations at the approach of winter, and taking shelter in the valleys; or, in case of a severe winter, visiting the estuaries.

They moult twice in the year, in June and November; in June, the males acquire the female plumage to a certain extent, but regain their proper dress at the second moult, and retain it during the breeding season. In a wild state, the mallard always pairs, and, during the period of incubation, the male, although taking no part in the process, always keeps in the neighborhood of the female; and it is singular that half-bred birds between the wild and tame varieties always exhibit the same habits, although the ordinary domestic drakes are polygamous, always endeavoring to get as many wives as they can. The nest is usually placed upon the ground among reeds and ledges near the water; sometimes in holes or hollow trees, but rarely among the branches. The eggs vary from about eight to fourteen in number, and the young are active from the moment of their exclusion, and soon take to the water, where they are as much at home as the old birds.

As the flesh of wild ducks is greatly valued, immense numbers are shot, or taken in other ways. In England, large numbers are captured by decoys, consisting of a piece of water situated in the midst of a quiet plantation, from which six semicircular canals are cut, which are roofed over with hoops, and covered in with netting. Into this vast trap the ducks are enticed by young ducks trained for the purpose.

THE DOMESTIC DUCK.

The duck should always find a place in the poultry-yard, provided that it can have access to water, even a small supply of which will suffice. They have been kept with success, and the ordinary duck fattened to the weight of eight pounds, with no further supply of water than that afforded by a large pool sunk in the ground. In a garden, ducks will do good service, voraciously consuming slops, frogs, and insects—nothing, indeed, coming amiss to them; not being scratchers, they do not, like other poultry, commit such a degree of mischief, in return, as to partially counterbalance their usefulness. A drake and two or three ducks cost little to maintain; and the only trouble they will give is, that if there is much extent of water or shrubbery about their home, they will lay and sit abroad, unless they are brought up every night, which should be done. They will otherwise drop their eggs carelessly here and there, or incubate in places where their eggs will be sucked by crows, and half their progeny destroyed by rats.

The duck is very prolific, and its egg is very much relished by some, having a rich piquancy of flavor, which gives it a decided superiority over the egg of the domestic fowl; and these qualities render it much in request with the pastry-cook and confectioner—three duck's eggs being equal in culinary value to six hen's eggs. The duck does not lay during the day, but generally in the night; exceptions, regulated by circumstances, will, of course, occasionally occur. While laying, it requires, as has been intimated, more attention than does the hen, until it is accustomed to resort to a regular nest for depositing its eggs; when, however, this is once effected, little care is needed beyond what has been indicated.

ROUEN DUCK.

The duck is a bad hatcher, being too fond of the water, and, consequently, too apt to allow her eggs to get cold; she will also, no matter what kind of weather it may be, bring the ducklings to the water the moment they break the shell—a practice always injurious, and frequently fatal; hence the very common practice of setting duck's eggs under hens.

There are several *varieties* of tame ducks; but their merits are more diverse in an ornamental than in a profitable point of view. Of *white* ducks, the best is the *Aylesbury*, with its unspotted, snowy plumage, and yellow legs and feet. It is large and excellent for the table, but not larger or better than several others. They are assiduous mothers and nurses, especially after the experience of two or three seasons. A much smaller race of white ducks is imported from Holland, useful only to the proprietors of extensive or secluded waters, as enticers of passing wild birds to alight and join their society. This variety has a yellow-orange bill; that of the Aylesbury should be flesh-colored. There is, also, the *white hook-billed* duck, with a bill monstrously curved downward—a Roman-nosed duck, in fact—with Jewish features, of a most grotesque and ludicrous appearance; the bill has some resemblance in its curvature to that of the Flamingo. White ducks, of course, make but a sorry figure in towns or dirty suburbs, or in any place where the means of washing themselves are scanty.

There are one or two pretty varieties, not very common; one of a *slate-gray*, or bluish dun, another of a *sandy-yellow;* there are also some with top-knots as compact and spherical as those of any Polish fowl, which rival the hook-billed in oddity What are termed the *white* Poland and the *black* Poland are crested; they breed early, and are excellent layers; the former are deemed the most desirable though the black are the larger.

Of *mottled* and *pied* sorts, there exists a great variety; black and white, bronze and white, lightly speckled, and many other mixtures. To this class belongs the *Rouen*—or Rhone, or Rohan, since each designation has been used—duck, which has been needlessly overpraised by interested dealers. This variety is highly esteemed by epicures; it is a prolific bird, and lays large eggs; its size is the criterion of its value. There is also a pied variety of the *Poland* ducks, a hybrid between the white and the black, the Beaver.

Another variety, known as the *Labrador*, the Buenos Ayres, or the black East Indian duck, is somewhat rare and highly esteemed by dealers. They are very beautiful birds. The feet, legs, and entire plumage should be black; a few white feathers will occasionally appear. The bill also is black, with a slight under-tinge of green. Not only the neck and back, but the larger feathers of the tail and wings are gilt with metallic green; the female also exhibits slight traces of the same decoration. On a sunshiny spring day, the effect of these glittering black ducks sporting in the blue water is very pleasing.

A peculiarity of this variety is, that they occasionally—that is, at the commencement of the season—lay black eggs; the color of those subsequently laid gradually fades to that of the common kinds. This singular appearance is not caused by any internal strain penetrating the whole thickness of the shell, but by an oily pigment, which may be scraped off with the nail. They lay, perhaps, a little later than other ducks, but are not more difficult to rear. Their voice, likewise, is said to differ slightly from that of other varieties; but they are far superior in having a high, wild-duck flavor and, if well kept, are in deserved repute as being excellent food when killed immediately from the pond, without any fattening.

Still another breed, known as the *Muscovy* duck, is a distinct species from the common duck; and the hybrid race will not, therefore, breed again between themselves, although they are capable of doing so with either of the species from the commixture of which they spring. This duck does not derive its name from having been brought from the country indicated, but from the flavor of its flesh, and should more properly be termed the *musk* duck, of which this name is but a corruption. It is easily distinguished by a red membrane surrounding the eyes, and covering the cheeks. Not being in esteem, on account of their peculiar odor, and the unpleasant flavor of their flesh, they are not worth breeding, unless to cross with the common varieties; in which case, the musk drake must be put to the common duck. This will produce a very large cross, while the opposite course will beget a very inferior one.

THE GOOSE.

THE WILD GOOSE. The goose belongs to the same family as the duck, but is classed with the genus *anser*. The *gray-leg* goose—a common wild goose of England—is by some regarded as the original of the domestic bird. It is thirty-five inches long; upper parts ash-brown and ash-gray; under parts white. This variety is migratory, proceeding to the Northern parts of Europe and Asia in summer, and to the South in winter.

The *Canada*, or Cravat goose, the wild goose of this country, is a fine species, forty inches long, often seen in spring and autumn in large, triangular flocks, high in air, and led by an old, experienced gander, who frequently utters a loud *honk*, equivalent, doubtless, to "All's well!" This sound often comes upon the ear at night, when the flock are invisible; and

WILD OR CANADA GOOSE.

it is frequently heard even in the daytime, seeming to come from the sky, the birds being beyond the reach of vision. Immense numbers of these noble birds are killed in Canada, as well as along our coasts, where they assemble in the autumn in large flocks, and remain till driven to more Southern climates by the season.

The Canada goose is capable of domestication, and, in spite of its original migratory habits—which it appears, in almost every instance, to forget in England—shows much more disposition for a truly domestic life than the swan; and it may be maintained in perfect health with very limited opportunities for bathing. They eat worms and soft insects, as well as grass and aquatic plants; with us, they do not breed until they are at least two years old, and so far approach the swan; like which bird, also, the male appears to be fit for reproduction at an eariier period than the female. Many writers speak highly of the half-bred Canada. They are, certainly, very large, and may merit approbation on the table; but with whatever other species the cross is made, they are hideously disgusting.

THE DOMESTIC GOOSE.

The goose is not mentioned in the Bible, but it was known to the ancient Egyptians, and is represented in numerous instances on their monuments, showing that it was anciently used for food, as in our own times. It was held sacred by the Romans, because it was said to have alarmed, by its cackling at night, the sentinels of the capitol, at the invasion of the Gauls, and thus to have saved the city. This was attributed by one of the Roman writers to its fine sense of smell, which enables them to perceive at a great distance the odor of the human race. The liver of this bird seems to have been a favorite morsel with epicures in all ages; and invention appears to have been active in exercising the means of increasing the volume of that organ. It is generally esteemed a foolish bird; yet it displays courage in defending its young, and instances of attachment and gratitude have shown that it is not deficient in sentiment. The value and usefulness of geese are scarcely calculable. The only damage which they do lies in the quantity of food which they consume; the only care they require is to be saved from starvation. All the fears and anxieties requisite to educate the turkey and prepare it for making a proper appearance at the table are with them unnecessary; grass by day, a dry bed at night, and a tolerably attentive mother, are all that is required. Roast goose, fatted to the point of repletion, is almost the only luxury that is not deemed an extravagance in an economical farm-house; for there are the feathers, to swell the stock of beds; there is the dripping, to enrich the dumpling or pudding; there are the giblets, for market or a pie; and there is the wholesome, solid, savory flesh for all parties interested.

They are accused by some of rendering the spots where they feed offensive to other stock; but the explanation is simple. A horse bites closer than an ox; a sheep goes nearer to the ground than a horse; but, after the sharpest shearing by sheep, the goose will polish up the tuft, and grow fat upon the remnants of others. Consequently, where geese are kept in great numbers on a small area, little will be left to maintain any other grass-eating creature. If, however, the pastures are not short, it will not be found that other grazing animals object to feeding either together with, or immediately after, a flock of geese.

The goose has the merit of being the earliest of poultry. In three months, or, about four, from leaving the egg, the birds ought to be fit for the feather-bed, the spit, and the fire. It is not only very early in its laying, but also very late. It often anticipates the spring in November, and, afterward, when spring really comes in March, it cannot resist its general influence. The autumnal eggs afford useful employment to turkeys and hens that choose to sit at unseasonable times; and the period of incubation is less tedious than that required for the eggs of some other birds.

The flight of the domestic goose is quite powerful enough, especially in young birds, to allow them to escape in that way, where they are so inclined. In the autumn, whole broods may be seen by early risers taking their morning flight, and circling in the air for matutinal exercise, just like pigeons, when first let out of their locker. The bird lives to a very great age, sometimes seventy years or more.

As to the origin of our domestic species, opinions differ. By some, as already remarked, the gray-leg is regarded as the parent stock; others consider it a mongrel, like the dunghill fowl, made up of several varieties, to each of which it occasionally shows more or less affinity; and yet others contend that it is not to be referred to any existing species. The latter assert that there is really but one variety of the domestic goose, individuals of which are found from entirely white plumage, through different degrees of patchedness with gray, to entirely gray coloring, except on the abdomen.

The domestic gander is polygamous, but he is not an indiscriminate libertine; he will rarely couple with females of any other species. Hybrid common geese are almost always produced by the union of a wild gander with a domestic goose, and not by the opposite. The ganders are generally, though not invariably, white, and are sometimes called Embden geese, from a town of Hanover. High feeding, care, and moderate warmth will induce a prolific habit, which becomes, in some measure, hereditary. The season of the year at which the young are hatched—and they may be reared at any season—influences their future size and development. After allowing for these causes of diversity, it is claimed that the domestic goose constitutes only one species or permanent variety.

THE BERNACLE GOOSE.

This bird is sometimes called the Barnacle goose; its name originates from the fact that it was formerly supposed to be bred from the shells so termed, which cling to wood in the sea. It is about twenty-five inches long, and is found wild in Europe, abundantly in the Baltic; and, occasionally, as it is said, in Hudson's Bay, on this continent.

This bird is one of those species in which the impulse of reproduction has at length overcome the sullenness of captivity, and instances of their breeding when in confinement have increased in frequency to such an extent that hopes are entertained of the continuance of that increase. The young so reared should be pinioned at the wrist as a precaution. They would probably stay at home contentedly, if unpinioned, until the approach of inclement weather, when they would be tempted to leave their usual haunts in search of marshes, unfrozen springs, mud-banks left by the tide, and the open sea; or they might be induced to join a flock of wild birds, instead of returning to their former quarters.

Broods of five, six, and seven have been reared; but they are generally valued as embellishments to ponds merely, their small size rendering them suitable even for a very limited pleasure-ground, and the variety being perhaps the prettiest geese that are thus employed. The lively combination of black, white, gray, and lavender, gives them the appearance of being in agreeable half-mourning. The female differs little from the male, being distinguished by voice and deportment more than by plumage. Their short bill, the moderate-sized webs of their feet, and their rounded proportions, indicate an affinity with the curious Cereopsis goose, which is found in considerable numbers on the seashore of Sucky Bay and Goose Island, at the southeastern point of Australia.

The number of eggs laid is six or seven, and the time of incubation is about a month; it being difficult to name the exact period, from the uncertainty respecting the precise hour when the process commences. They are steady sitters. The young are lively and active little creatures, running hither and thither, and tugging at the blades of grass. Their ground color is of a dirty white; their legs, feet, eyes, and short stump of a bill, are black; they have a gray spot on the crown of the head, gray patches on the back and wings, and a yellowish tinge about the forepart of the head. The old birds are very gentle in their disposition and habits, and are less noisy than most geese.

The service they may render as weed-eaters is important, though their size alone precludes any comparison of them, in this respect, with the swan. Their favorite feeding-grounds are extensive flats, partially inundated by the higher tides; and their breeding may perhaps best be promoted by their being furnished with a little sea-weed during winter and early spring; a few shrimps, or small muscles, would probably not be unacceptable. A single pair is more likely to breed than if they are congregated in larger numbers.

THE BREMEN GOOSE.

The Bremen geese—so called from the place whence they were originally imported, though some term them Embden

A BREMEN GOOSE.

geese—have been bred in this country, pure, and to a feather, since 1821; no single instance having occurred in which the slightest deterioration of character could be observed. The produce has invariably been of the purest white; the bill, legs, and feet being of a beautiful yellow.

The flesh of this goose does not partake of that dry character which belongs to other and more common kinds, but is as tender and juicy as the flesh of a wild fowl; it shrinks less in cooking than that of any other fowl. Some pronounce its flesh equal if not superior to that of the canvas-back duck.

They likewise sit and hatch with more certainty than common barn-yard geese; will weigh nearly, and in some instances quite, twice the weight—the full-blood weighing twenty pounds and upward; they have double the quantity of feathers; and never fly.

THE BRENT GOOSE.

This is a small species, twenty-one inches long, common in a wild state, in both Europe and America. On our coast, it is a favorite game-bird, and known by the name of *Brant*. It is easily tamed, and is said to have produced young in captivity, though no details have been furnished.

This and the Sandwich Island goose are the smallest of their tribe yet introduced to our aquatic aviaries. Their almost uniform color of leaden black, and their compactness of form, make them a striking feature in the scene, though they cannot be compared in beauty with many other water-fowl. There is so little difference in the sexes that it is not easy to distinguish them. Their chief merit rests in their fondness for water-weeds, in which respect they appear to be second only to the swan. They are quiet, gentle, and harmless in captivity. Some praise their flesh, while others pronounce it fishy, strong, and oily; they may, however, be converted into tolerable meat by being skinned and baked in a pie.

THE CHINA GOOSE.

This bird figures under a variety of *aliases:* Knob goose, Hong Kong goose, Asiatic goose, Swan goose, Chinese Swan,

CHINA OR HONG KONG GOOSE.

Guinea goose, Polish goose, Muscovy goose, and, probably, others.

There is something in the aspect of this creature—in the dark-brown stripe down its neck, its small, bright eye, its harsh voice, its ceremonious strut, and its affectation of seldom being in a hurry—which seems to say that it came from

China. If so, it has no doubt been domesticated for many hundred years, perhaps as long as the pea fowl or the common fowl. They may be made to lay a large number of eggs by an increased supply of nourishing food. If liberally furnished with oats, boiled rice, etc., the China goose will, in the spring, lay from twenty to thirty eggs before she begins to sit, and again in the autumn, after her moult, from ten to fifteen more. Another peculiarity is their deficient power of flight, compared with other geese, owing to the larger proportionate size of their bodies. Indeed, of all geese, this is the worst flyer; there is no occasion to pinion them; the common domestic goose flies much more strongly.

The prevailing color of their plumage is brown, comparable to the color of wheat. The different shades are very harmoniously blended, and are well relieved by the black tuberculated bill, and the pure white of the abdomen. Their movements on the water are graceful and swan-like. Slight variations occur in the color of the feet and legs, some having them of a dull orange, others black; a delicate fringe of minute white feathers is occasionally seen at the base of the bill. These peculiarities are hereditarily transmitted.

The male is almost as much disproportionately larger than the female as the Musk drake is in comparison with his mate. He is much inclined to libertine wanderings, without, however, neglecting proper attention at home. If there is any other gander on the premises, a disagreement is sure to result. Both male and female are, perhaps, the noisiest of all geese; at night, the least footfall or motion in their neighborhood is sufficient to call forth their clangor and resonant trumpetings.

The eggs are somewhat less than those of the domestic kind,

of a short oval, with a smooth, thick shell, white, but slightly tinged with yellow at the smaller end. The goslings, when first hatched, are usually very strong. They are of a dirty green, like the color produced by mixing India-ink and yellow ochre, with darker patches here and there. The legs and feet are lead-color, but afterward change to a dull red. With good pasturage, they require no farther attention than that bestowed by their parents. After a time, a little grain will strengthen and forward them. If well fed, they come to maturity very rapidly; in between three and four months from the time of leaving the shell, they will be full-grown and ready for food. They do not bear being shut up to fatten so well as common geese, and, therefore, those destined for the table are the better for profuse hand-feeding. Their flesh is well-flavored, short, and tender; their eggs, excellent for cooking purposes

They are said to be a short-lived species; the ganders, at least, not lasting more than ten or a dozen years. Hybrids between them and the common goose are prolific with the latter; the second and third cross is much prized by some farmers, particularly for their ganders; and in many flocks the blood of the China goose may be traced oftentimes by the more erect gait of the birds, accompanied by a faint stripe down the back of the neck. With the White-grented goose they also breed freely.

The White-China. These are larger than the preceding, and apparently more terrestrial in their habits; the knob on the head is not only of greater proportion, but of a different shape. It is of a spotless, pure white—though a very few gray feathers occasionally appear—more swan-like than the brown, with a bright orange-colored bill, and a large knot of the same color at its base. It is particularly beautiful, either in or out of the water, its neck being long, slender, and gracefully arched when swimming. It breeds three or four times in the season; the egg is quite small for the size of the bird, being not more than half the size of that of the common goose.

In many instances, efforts to obtain young from their eggs have been unsuccessful; but if the female is supplied with the eggs of the common goose, she invariably hatches and rears the goslings. They sit remarkably well, never showing themselves out of the nest by day; but, possibly, they may leave the nest too long in the cold of the night. Some think that a quiet lake is more to their taste than a rapid running stream, and more conducive to the fecundity of their eggs. It is also believed by many that, under favorable circumstances, they would be very prolific.

THE EGYPTIAN GOOSE.

This species is bred to a certain extent in this country. It is a most stately and rich bird, reminding one of the solemn antiquity of the Nile, with its gorgeous mantle of golden hues and its long history.

It is dark red round the eyes; red ring round the neck; white bill; neck and breast light fawn-gray; a maroon star on the breast; belly red and gray; half of the wing-feathers rich black, the other part of them pure white; black bar running across the centre, back light-red, growing dark-red toward the tail; the tail a deep black.

They are very prolific, bringing off three broods a year, from eight to twelve each time; their weight is about eight pounds each.

THE JAVA GOOSE.

The gander of this species is white, with head and half the neck light-fawn ; red tubercle at the root of the bill ; larger than the common goose, and longer in the body ; walks erect, standing as high as the China goose, the female appearing to carry two pouches, or egg-bags, under the belly.

It is very prolific ; and the meat is of fine flavor.

THE TOULOUSE GOOSE.

This bird is said to have been originally imported from the Mediterranean ; and is known also by the names Mediterranean goose, and Pyrenean goose. It is chiefly remarkable for its vast size, in which respect it surpasses all others.

Its prevailing color is a slaty blue, marked with brown bars, and occasionally relieved with black ; the head, neck, as far as the beginning of the breast, and the back of the neck, as far as the shoulders, of a dark-brown ; the breast slaty-blue ; the belly is white, in common with the under surface of the tail ; the bill is orange-red, and the feet flesh-color.

In habit, the Toulouse goose resembles his congeners, but seems to possess a milder and more tractable disposition, which greatly conduces to the chance of his early fattening, and that, too, at a little cost. The curl of the plumage on the neck comes closer to the head than that on common geese, and the abdominal pouch, which, in other varieties, is an accompaniment of age, exists from the shell. The flesh is said to be tender and well-flavored.

Some pronounce this bird the unmixed and immediate descendant of the Gray-leg ; while others assert that it is only the common domestic, enlarged by early hatching, very liberal feeding during youth, fine climate, and, perhaps, by age, and style them grenadier individuals of the domestic goose—nothing more.

THE WHITE-FRONTED GOOSE.

In its wild state, the White-fronted or Laughing goose is twenty-seven inches long, and found in great numbers in Europe and in the North American Fur countries, but rare along our coasts.

When domesticated, it belongs to the class of birds which are restrained from resuming their original wild habits more by the influence of local and personal attachment than from any love which they seem to have for the comforts of domestication; which may be trusted with their entire liberty, or nearly so, but require an eye to be kept on them from time to time, lest they stray away and assume an independent condition. The white-fronted goose well deserves the patronage of those who have even a small piece of grass.

The first impression of every one, upon seeing this species in confinement, would be that it could not be trusted with liberty; and the sight of it exercising its wings at its first escape would make its owner despair of recovering it. This is not, however, the case. By no great amount of care and attention, they will manifest such a degree of confidence and attachment as to remove all hesitation as to the future; and they may be regarded as patterns of all that is valuable in anserine nature—gentle, affectionate, cheerful, hardy, useful, and self-dependent. The gander is an attentive parent, but not a faithful spouse.

The eggs are smaller than those of the common goose, pure

white, and of a very long oval; the shell is also thinner than in most others; the flesh is excellent.

Having completed the enumeration and description of the varieties of poultry, it will, perhaps, be appropriate to give some account, before proceeding to the next general division of the subject, of the structure, or anatomy, so to speak, of

THE EGG.

In a laying hen may be found, upon opening the body, what is called the *ovarium*—a cluster of rudimental eggs, of different sizes, from very minute points up to shapes of easily-distinguished forms. These rudimental eggs have as yet no shell or white, these being exhibited in a different stage of development; but consist wholly of *yolk*, on the surface of which the germ of the future chicken lies. The yolk and the germ are enveloped by a very thin membrane.

When the rudimental egg, still attached to the ovarium, becomes longer and larger, and arrives at a certain size, either its own weight, or some other efficient cause, detaches it from the cluster, and makes it fall into a sort of funnel, leading to a pipe, which is termed the *oviduct*.

Here the yolk of the rudimental egg, hitherto imperfectly formed, puts on its mature appearance of a thick yellow fluid; while the rudimental chick or embryo, lying on the surface opposite to that by which it had been attached to the ovarium, is white, and somewhat like paste.

The white, or *albumen*, of the egg now becomes diffused around the yolk, being secreted from the blood vessels of the egg-pipe, or oviduct, in the form of a thin, glassy fluid; and it is prevented from mixing with the yolk and the embryo chicken by the thin membrane which surrounded them before they were detached from the egg-cluster, while it is strengthened by a second and stronger membrane, formed around the first, immediately after falling into the oviduct. This second membrane, enveloping the yolk of the germ of the chicken, is thickest at the two ends, having what may be termed bulgings, termed *chalazes* by anatomists; these bulgings of the second membrane pass quite through the white at the ends, and being thus, as it were, embedded in the white, they keep the inclosed yolk and germ somewhat in a fixed position, preventing them from rolling about within the egg when it is moved.

The white of the egg being thus formed, a third membrane, or, rather, a double membrane, much stronger than either of the first two, is formed around it, becoming attached to the chalazes of the second membrane, and tending still more to keep all the parts in their relative positions.

During the progress of these several formations, the egg gradually advances about half way along the oviduct. It is still, however, destitute of the shell, which begins to be formed by a process similar to the formation of the shell of a snail, as soon as the outer layer of the third membrane has been completed. When the shell is fully formed, the egg continues to advance along the oviduct, till the hen goes to her nest and lays it.

From ill health, or accidents, eggs are sometimes excluded from the oviducts before the shell has begun to be formed, and in this state they are popularly called *wind-eggs*.

Reckoning, then, from the shell inward, there are *six* different envelopes, of which one only could be detected before the descent of the egg into the oviduct: the shell; the external layer of the membrane lining the shell; the internal layer of same lining; the white, composed of a thinner liquid on the outside, and a thicker and more yellowish liquid on the inside; the bulgings, or chalaziferus membrane; and the proper membrane.

One important part of the egg is the *air-bag*, placed at the larger end, between the shell and its lining membrane. This is about the size of the eye of a small bird in new-laid eggs, but is increased as much as ten times in the process of hatching. The air bag is of such great importance to the development of the chicken—probably by supplying it with a limited atmosphere of oxygen—that, if the blunt end of an egg be pierced with the point of the smallest needle, the egg cannot be hatched.

Instead of one rudimental egg falling from the ovarium, two may be detected, and will, of course, be inclosed in the same shell, when the egg will be double-yolked. The eggs of a goose have, in some instances, contained even three yolks. If the double-yolked eggs be hatched, they will rarely produce two separate chickens, but, more commonly, monstrosities—chickens with two heads, and the like.

The *shell* of an egg, chemically speaking, consists chiefly of carbonate of lime, similar to chalk, with a small quantity of phosphate of lime, and animal mucus. When burnt, the animal matter and the carbonic acid gas of the carbonate of lime are separated; the first being reduced to ashes, or animal charcoal, while the second is dissipated, leaving the decarbonized lime mixed with a little phosphate of lime.

The *white* of the egg is without taste or smell, of a viscid, glairy consistence, readily dissolving in water, coagulable by acids, by spirits of wine, and by a temperature of one hundred and sixty-five degrees, Fahrenheit. If it has once been coagulated, it is no longer soluble in either cold or hot water, and acquires a slight insipid taste. It is composed of eighty parts of water, fifteen and a half parts of albumen, and four and a half parts of mucus; besides giving traces of soda, benzoic acid, and sulphuretted hydrogen gas. The latter, on an egg being eaten on a silver spoon, stains the spoon of a blackish purple, by combining with the silver, and forming sulphuret of silver.

The white of the egg is a very feeble conductor of heat, retarding its escape; and preventing its entrance to the yolk; a providential contrivance, not merely to prevent speedy fermentation and corruption, but to arrest the fatal chills, which might occur in hatching, when the mother hen leaves her eggs, from time to time, in search of food. Eels and other fish which can live long out of water, secrete a similar viscid substance on the surface of their bodies, furnished to them, doubtless, for a similar purpose.

The *yolk* has an insipid, bland, oily taste; and, when agitated with water, forms a milky emulsion. If it is long boiled it becomes a granular, friable solid, yielding upon expression, a yellow, insipid, fixed oil. It consists, chemically, of water, oil, albumen, and gelatine. In proportion to the quantity of albumen, the egg boils hard.

The weight of the eggs of the domestic fowl varies materially; in some breeds, averaging thirty-three ounces per dozen, in others, but fourteen and a half ounces. A fair average weight for a dozen is twenty-two and a half ounces. Yellow, mahogany, and salmon-colored eggs are generally richer than white ones, containing, as they do, a larger quantity of yolk. These are generally preferred for culinary purposes; while the latter, containing an excess of albumen, are preferred for boiling, etc., for the table.

Breeding. Good fowls are very profitable in the keeping of intelligent breeders. It is stated, by those most competent to express the opinion, that four acres of land, devoted to the rearing of the best varieties of poultry, will, at ordinary prices, be quite as productive as a farm of one hundred and fifty acres cultivated in the usual way. The eggs of the common and cheaper kinds which might be used for incubators and nurses, would pay—or could be made to pay, if properly preserved, and sold at the right time—all expenses of feed, etc.; while good capons of the larger breeds will bring, in any of our larger markets, from three to five dollars per pair, and early spring chickens from twenty to twenty-five cents per pound.

To make poultry profitable, then, it is only necessary that the better kinds be bred from, that suitable places be provided for them, that they be properly fed, and carefully and intelligently managed. These requirements are too rarely complied with, in every respect, to enable a correct opinion to be formed as to what may be made out of poultry under the most favorable circumstances.

A few general principles, well-understood and faithfully applied, will prove of great value. By "in-and-in breeding" is meant commerce between individuals of the same brood, or brother and sister, so to speak; by "close breeding," commerce between the parent and his offspring, in whatever degree.

Crossing the breed. To insure successful and beneficial crossing of distinct breeds, in order to produce a new and valuable variety, the breeder must have an accurate knowledge of the laws of procreation, and the varied influences of parents upon their offspring. All the breeds in this country are crosses, produced either by accident or design. Crossing does not necessarily produce a breed; but it always produces a variety, and that variety becomes a breed only where there is a sufficiency of stamina to make a distinctive race, and continue a progeny with the uniform or leading characteristics of its progenitors.

High breeding. When uniformity of plumage can be effected in mixed breeds or varieties without a resort to in-and-in, or close breeding, and without sacrificing the health and vigor of the race, it is desirable; and, in many instances, it can be accomplished in a satisfactory manner. What are called highly-bred fowls are, however, too often the deteriorated offspring of progenitors far below the original stock. Genuine high breeding consists in the selection of parent stock of the same race, perfect in all the general characteristics, and *of remote consanguinity.* This should be resorted to periodically, in order to secure the best results.

If a race is *pure*—that is, if the species or variety is absolutely distinct and unsophisticated—the progeny resembles the progenitors in almost every respect. The mixture of races, where the consanguinity is remote, is productive of decided benefits.

To illustrate, in the case of fowls: when the blood is *unmixed*—as with the Guelderlands, and some others—the offspring, *in all respects*, resemble their parents; in plumage, general habits, form, outline, etc. In this case, they look almost identically the same. But when the blood is *mixed*—as with the Cochin Chinas, and many others—the plumage will vary widely, or slightly, according to circumstances, though many or most of the general characteristics may remain the same. The close breeding, to which many resort for the purpose of procuring uniformity, generally results in an absolute deterioration of the race in important respects.

In some cases, close breeding—and, occasionally, in-and-in—seems to be in accordance with the laws of Nature; as with the wild turkey, which, in its natural state, resorts to these modes of breeding; and yet the race does not change in appearance or degenerate. The reason is that the breed is pure. In comparing any number of these birds, not the least dissimilarity is discoverable; they all look alike, as they always have, and always will. They are changed, or deteriorated, only by crossing or confinement.

Most breeds of the hen kind degenerate rapidly from close, or in-and-in breeding, because they are not perfect of their kind; that is, the breed is not pure, but of mixed blood; and in such objectionable breeding, the race degenerates just in proportion as the breed is imperfect, or impure. The perfect Guelderland will admit of these modes of breeding, for a great length of time, without deterioration; but the impure or mixed will rapidly degenerate. This is also true of all breeds, wherein the characteristic marks are uniform and confirmed, showing perfection in the race.

As a general rule, however, close and in-and-in breeding should be carefully avoided where the race is not absolutely perfect, if it is desired to improve the breed; and as all the breeds of this kind of fowls are of mixed blood, the danger of such breeding is greater or less, in exact proportion as the distinctive characteristics are variant or fixed; and the danger still increases if the breed is composed of strains of blood greatly dissimilar, or of races widely differing in the conformation or general habits.

Preserving the distinctive breeds. As to the time when the different breeds of hens should be separated in the spring, in order to preserve the breed pure, the most ample experience indicates that if the eggs be preserved and set after a separation of *two days*, the breed will be perfect, the offspring having all the characteristics or distinctive marks.

When a valuable breed is produced, either by accident or design, it should be preserved, and the subsequent breeding should continue from that stock; otherwise, there is no certainty of the purity of the blood of the new breed, for it does not follow that a different parentage, though of the same name or original breed precisely, will produce the same new breed, or any thing resembling it. The Dorking fowl, for instance, was originally produced by crossing the Great Malay with the English Game, as an accident; but it by no means follows that Dorkings are the uniform, or even the common result of such a cross, for hundreds of similar experiments have proved unsuccessful. The breeding, therefore, to be pure-blooded, must continue from the stock originally produced by accident; and as such breeding produces the leading characteristics of the race with great uniformity, the genuineness of the breed cannot be doubted.

In order to produce a good cross, the parentage should be healthy, and from healthy races, not materially dissimilar in their general habits. The *size of the leg* should always be looked to, in order to judge accurately as to purity of blood. If the leg is large for the breed—that is, if larger than the generality of the same breed—the purity of the blood, the fineness of the flesh, and most of the other valuable qualities, can be relied on; but, if the legs are smaller than most others of the same breed, the fowl is spurious, and of deteriorated blood. The fifth toe and feathered legs of some breeds were originally the result of accident; but by long and careful breeding, they have become incorporated into the nature of certain races of general, though not universal or essential, requisites. When a fowl exhibits any special marks indicative of all the races or breeds from which the cross originated, it is a sure evidence of extraordinary purity of blood, and of the superior excellence of the race. The best fowls of the race should always be selected for crossing or general breeding; otherwise the breeds will degenerate.

The *quality*—that is, the fineness, juiciness, and richness of flavor—of the flesh of domestic fowls is of much more importance than their size. All coarse-meated fowls should, therefore, be rejected, no matter how large they may be. There is no difficulty in discriminating between coarse and fine fowls at any time. In the case of chickens, if the down is straight and stands out, and the body and limbs are loosely joined, the meat is coarse; but if the down is glossy, and lies close to the body, and the body and limbs are compactly formed, the meat is fine; and when grown, if the fowl is light in weight, in proportion to its size, the flesh is coarse; but if heavy, the flesh is fine.

There is also a *fitness* in the quality of the flesh; for, if the meat is fine, the bones are fine, and the feathers are fine; and the converse holds true. If the flesh is fine, it is juicy and richly flavored; if coarse, it is dry, fibrous, and insipid.

The *color of the legs*, too, is quite material in judging of the quality of fowls. All other things being equal, dark-legged fowls have the finest flesh, and are most hardy. Turkeys, which have the finest flesh of any fowl of their size, have black legs; the game-cock, likewise, which is universally acknowledged to be the finest-fleshed of any of the domestic fowls, except the Wild Indian fowl of Calcutta, has dark legs. It does not, however, of necessity follow that all dark-legged fowls are fine, or that all yellow or white-legged ones are coarse, since much depends upon the breed; but it is true that the darkest leg which pertains to the breed indicates the finest fowl.

The *color of the feathers*, also, has more or less to do with the quality of the fowl. Some breeds have a much more brilliant plumage than others; but when brilliancy of plumage is here spoken of, it is to be understood in comparison with others of the same breed. If, therefore, a fowl is selected of rich and glossy plumage, when compared with others of the same breed, the legs will be dark of the kind, and the quality of the bird will excel.

The *best* breeding is to cross or mix the races; this process improves the breeds, in all respects. When the object in view is to perpetuate distinct varieties of uncontaminated blood, the first requisite is to procure fowls known to be of pure blood, and possessing all the necessary characteristics of their kind. Labor is lost, unless the fowl selected is a perfect specimen of the variety; for whatever imperfection exists is likely to be perpetuated in the progeny. Regard should be had to plumage, size, and form, in making a selection either of a cock or a pullet; and those are preferable which are hatched earliest in the year. The *age* of the fowls is a matter of considerable importance; and, though it is true that a pullet will lay the greatest number of eggs in her first year, yet it is believed that the chickens which are hatched from the second year's eggs are more vigorous and hardy. Old hens are generally preferred to pullets as sitters, on account of their more sedate and matronly character. A young cock, though more active in his earliest days, and likely to bestow his attention on the hens with less reserve, is not, however, best for use in keeping up a breed. The eggs impregnated by him after his first season are likely to produce the strongest chickens. It is an error to suppose—as is often represented—

that his procreative power is decayed or vitiated after three or four years. On the contrary, a healthy, vigorous cock, if not allowed to walk with too many hens, may be valuable and useful in the poultry-yard for a longer time.

An error is often committed by assigning too many hens to one cock; and the result is a weakly and otherwise deteriorated progeny. Not more than *five* hens should be allowed to associate with a single cock, when the quality of the breed is a matter of interest. *Three*, indeed, would be the better number for restriction; but five is the farthest limit which can be safely assigned.

Most persons, in obtaining a single vigorous cock and hen of a desirable variety, find their anticipations more than realized in the production of a fine progeny. The plumage is brilliant, and the chickens are of increased size, and remarkably strong and healthy. This desirable state of things continues so long as the cock is restricted to a small number of hens; but as soon as his harem is enlarged, different effects are manifested, and a deterioration in the stock is clearly observable—attributable, not to close-breeding, but to the increased disproportion of the females to the male, and the consequent overtasking of his powers.

In breeding-time, great cleanliness should be preserved in the lodgings of the fowls, and the quantity and quality of food should be attended to. They should not be suffered to feed to repletion, and such kinds of food as are most nutritious should be carefully provided. Variety of food is essential; and a proper proportion of animal and green food should be given with their usual fare. Suitable arrangements should, of course, be made to prevent any intermixture of breeds. A constant vigilance in this respect is the price of success; and when all proper precautions are taken, the breeder may be perfectly secure that his anticipations will be realized.

SELECTION OF STOCK.

The habits of the domestic fowl, in a wild state, are too little known to ascertain whether the cocks always associate with the hens, or only occasionally. Though hens will lay some eggs without pairing, as this is not natural, the number will, for the most part, be less, and the laying uncertain; it is, therefore, indispensable to attend to the laws of Nature in this respect.

The number of hens to be allowed to one cock should vary with the object in view. The limit for valuable breeding purposes has already been indicated. If profit is sought for, in the production of eggs alone, one cock—if a stout, young, and lively bird—may have as many as twenty-four hens.

FIGHTING COCKS.

The choice of a cock is a very important thing. He is considered to have every requisite quality when he is of a good middling size; carries his head high; has a quick, animated look; a strong and shrill voice; a fine red comb, shining as if varnished; wattles of a large size, and of the same color as the comb; the breast broad; the wings strong; the plumage black or of an obscure red; the thighs very muscular; the legs thick, and furnished with strong spurs; and the claws rather bent and sharply pointed. He ought, also, to be free in his motions, to crow frequently, and to scratch the ground often in search of worms, not so much for himself as to treat his hens. He ought, withal, to be brisk, spirited, ardent, and ready in caressing the hens; quick in defending them, attentive in soliciting them to eat, in keeping them together, and in assembling them at night.

In breeding *game cocks*, the qualities required are every mark of perfect health, such as a ruddy complexion; the feathers close, short, and not feeling cold or dry; the flesh firm and compact; and a full breast, betokening good lungs; a tapering and thinness behind. He should be full in the girth, well coupled, lofty and aspiring, with a good thigh, the beam of his leg very strong, the eye large and vivid, and the beak strong, crooked, and thick at the base.

A cock is in his prime at two years old; though cocks are sometimes so precocious as to show every mark of full vigor at four months, while others of the same brood do not appear in that state for several months afterward. When marks of declining vigor are perceived, the cock must be displaced, to make way for a successor, which should be chosen from among the finest and bravest of the supernumerary young cocks, that ought to be reared for this special purpose.

The change of cocks is of much importance, and is frequently very troublesome to manage; for peace does not long subsist between them when they hold a divided dominion in the poultry-yard, since they are all actuated by a restless, jealous, hasty, fiery, ardent disposition; and hence their quarrels become no less frequent than sanguinary. A battle soon succeeds to provocation or affront. The two opponents face each other, their feathers bristling up, their necks stretched out, their heads low, and their beaks ready for the onslaught. They observe each other in silence, with fixed and sparkling eyes. On the least motion of either, they stand stiffly up, and rush furiously forward, dashing at each other with beak and spur in repeated sallies, till the more powerful or the more adroit has grievously torn the comb and wattles of his adversary, has thrown him down by the heavy stroke of his wings, or has stabbed him with his spurs.

In *the choice of a hen* for sitting, a large bird should be selected, with large, wide-spreading wings. Though large, she must not, however, be heavy nor leggy. No one of judgment would sit a Malay; as, in such case, not only would many eggs remain uncovered, but many, also, would be trampled upon and broken. Elderly hens will be more willing to sit than young and giddy pullets.

After the common hen, which, on account of her fecundity, is deservedly esteemed, the tufted hens may be justly ranked; particularly from being more delicate eating, because she fattens more readily, on account of laying less. The large breed, though less prolific, is preferable in rearing chickens for the market, or for making capons. With regard to these three kinds, the general opinion of breeders is, that the first is more prolific in the number of eggs, while the others produce larger chickens, which bring good prices.

The Spanish fowl are not generally good sitters, but are excellent layers; the Dorkings reverse the order, being better sitters than layers. These qualities will be found to extend pretty generally to hens partaking of the prevailing colors of these two varieties; the black being usually the best layers, and but careless or indifferent sitters, while gray or checkered hens are the best that can be produced.

FEEDING.

Experiments have demonstrated that what may be called the gastric juice in fowls has not sufficient power to dissolve their food, without the aid of the grinding action of the gizzard. Before the food is prepared for digestion, therefore, the grains must be subjected to a triturating process; and such as are not sufficiently bruised in this manner, before passing into the gizzard, are there reduced to the proper state, by its natural action. The action of the gizzard is, in this respect, mechanical; this organ serving as a mill to grind the food to pieces, and then, by means of its powerful muscles, pressing it gradually into the intestines, in the form of pulp. The power of this organ is said to be sufficient to pulverize hollow globules of glass in a very short time, and solid masses of the same substance in a few weeks. The rapidity of this process seems to be proportionate, generally, to the size of the bird. A chicken, for example, breaks up such substances as are received into its stomach less readily than the capon; while a goose performs the same operation sooner than either. Needles, and even lancets, given to turkeys, have been broken in pieces and voided, without any apparent injury to the stomach. The reason, undoubtedly, is, that the larger species of birds have thicker and more powerful organs of digestion.

It has long been the general opinion that, from some deficiency in the digestive apparatus, fowls are obliged to resort to the use of stones and gravel, in order to enable them to dispose of the food which they consume. Some have supposed that the use of these stones is to sheath the gizzard, in order to fit it to break into smaller fragments the hard, angular substances which might be swallowed; they have also been considered to have a medicinal effect; others have imagined that they acted as absorbents for undue quantities of acids in the stomach, or as stimulants to digestion; while it has even been gravely asserted that they contribute directly to nutrition.

Repeated experiments, however, have established that pebbles are not at all necessary to the trituration of the hardest kinds of substances which can be introduced into their stomachs; and, of course, the usual food of fowls can be bruised without their aid. They do, however, serve a useful auxiliary purpose. When put in motion by the muscles, they are capable of producing some effects upon the contents of the stomach; thus assisting to grind down the grain, and separating its parts, the digestive fluid, or gastric juice, comes more readily in contact with it.

VARIETIES OF FOOD. Fowls about a poultry-yard can usually pick up a portion of their subsistence, and, under favorable circumstances, the largest portion. When so situated, the keeping of poultry pays decidedly the best. The support even of poultry not designed for fattening should not, however, be made to depend entirely upon such precarious resources. Fowls should be fed with punctuality, faithfulness, and discretion.

They are fond of all sorts of grain—such as Indian corn, wheat, oats, rye, buckwheat, barley, millet, etc.; but their particular preferences are not so likely to guide in the selection of their food, as the consideration of what is most economical, and easiest to be procured on the part of their owner. They will readily eat most kinds of vegetables in their green state, both cooked and raw. They likewise manifest an inclination for animal food—such as blood, fish, and flesh—whether raw or otherwise; and seem by no means averse to feeding on their own species. Insects, worms, and snails they will take with avidity.

It is usual to give to domestic fowls a quantity of grain once, at least, daily; but, commonly, in less quantity than they would consume, if unrestricted. They feed with great voracity; but their apparent greediness is not the criterion by which the possibility of satisfying them is to be judged. Moderate quantities of food will suffice; and the amount consumed will usually be proportioned to the size of the individuals. Whatever is cheapest, at any given time, may be given, without regard to any other considerations. Different circumstances and different seasons may occasion a variation in their appetite; but a gill of grain is, generally speaking, about the usual daily portion. Some very voracious fowls, of the largest size, will need the allowance of a third of a pint each day.

Wheat is the most nutritive of cereal grains—with, perhaps, the exception of rice—as an article of human food. It is, therefore, natural to suppose that it is the best for fowls; and the avidity with which they eat it would induce the conclusion that they would eat more of this than of any other grain. Yet it appears that when fowls have as much wheat as they can consume, they will eat about a fourth part less than of oats, barley, or buckwheat; the largest quantity of wheat eaten by a fowl in one day being, according to several experiments, about three-sixteenths of a pint. The difference in bulk is, however, compensated by the difference in weight, these three-sixteenths of wheat weighing more than one-fourth of a pint of oats. The difference in weight is not, in every instance, the reason why a fowl is satisfied with a larger or smaller measure of one sort than another. *Rye* weighs less than wheat; but still a fowl will be satisfied with half the quantity of this grain. *Indian corn* ranks intermediately between wheat and rye; five-fourths of a pint of Indian corn with fowls being found, by experiment, equal to six-fourths of wheat, and three-fourths of rye.

In estimating the quantity of grain daily consumed by the common fowl, it is wise to use data a little above than below the average. It may, therefore, safely be said that a fowl of the common size, having free access to as much as can be eaten through the day, will consume, day by day, of oats, buckwheat, or barley, one-fourth of a pint; of wheat, three-sixteenths; of Indian corn, five thirty-seconds; and of rye, three thirty-seconds.

It has been conclusively settled, by experiments instituted to that end, that there is the best economy in feeding poultry with *boiled* grain rather than with dry, in every case where Indian corn, barley, and wheat can be procured. The expense of fuel, and the additional trouble incident to the process of cooking, are inconsiderable in comparison with the advantages derived. Where oats, buckwheat, or rye are used, boiling is useless, when profit is concerned.

BRAN. It is an erroneous notion that money can be saved

by feeding bran to fowls; since, then, so little of the farina of the grain remains in it, that the nourishment derived from its use is hardly worth mentioning. When boiled, as it always must be, its bulk is but slightly increased. Two measures of dry bran, mixed with water, are equal to but three-fifths of a measure of dry barley.

MILLET. This is recommended as excellent food for young chickens. Fowls always prefer it raw; though, as its bulk is increased one-half by boiling, it is doubtless more economical to feed it cooked.

RICE. Fowls are especially fond of this food, although they soon lose their relish for it when allowed to have it at their discretion. It should always be boiled; but its expense puts it out of the question as a daily diet. When used continuously, it should always be mixed with some substance containing less nutritive matter, in order that the appetite may not be cloyed by it.

POTATOES. These are very nutritious, and are usually acceptable to fowls, when properly prepared. When raw, or in a cold state, they appear to dislike them; they should, therefore, be boiled and given when moderately hot; when very hot, it is said that fowls will injure themselves by eating them, and burning their mouths. They should also be broken into pieces of convenient size; otherwise, they will be avoided. Occasionally raw pieces of potato will be devoured; but fowls cannot be said to be fond of the root in this state. The same remark applies to most other roots, especially to *carrots* and *parsnips;* these should always be prepared, in order to be wholesome and palatable. Fowls should never be confined to a root diet, in any case; but such food should be mingled or alternated with a sufficient quantity of grain.

Green food. Indulgence in this kind of diet is absolutely necessary to the health of fowls, and is also advantageous in an economical point of view. The more delicate kinds of green vegetables are eaten with the utmost avidity; all succulent weeds, grass, and the leaves of trees and shrubs will also be consumed. If hens have green plots to graze in during the day, the expense of their keeping will be reduced one-half. All the refuse of the kitchen, of a vegetable nature, should be freely thrown into the poultry-yard.

Green food, however, will not answer for an exclusive diet. Experiment has shown that fowls fed with this food alone for a few days together exhibit severe symptoms of relaxation of the bowels; and, after the lapse of eight or nine days, their combs become pale and livid, which is the same indication of disease in them that paleness of the lips is in the human species.

Earth-worms. These are regarded as delicacies by the inhabitants of the poultry-yard; and the individual who is fortunate enough to capture one is often forced to undergo a severe ordeal in order to retain his captive. Earth-worms are more plentiful in moist land, such as pastures, etc., than in that which is cultivated; in gardens, also, they exist in vast numbers. When it is desirable to take worms in quantities, it is only necessary to thrust a stake or three-pronged fork into the ground, to the depth of about a foot, and to move it suddenly backward and forward, in order to shake the soil all around; the worms are instinctively terrified by any motion in the ground, and, when disturbed, hasten to the surface.

It is advisable to store worms, on account of the trouble and difficulty of making frequent collections. They may be placed in casks, filled one-third full with earth, in quantities at least equal in bulk to the earth. The earth should be sprinkled occasionally, to prevent it from becoming too dry. Care should, however, be exercised that the earth does not become too moist; since, in such an event, the worms will perish. In rainy weather, the casks should be protected with a covering.

ANIMAL FOOD. Fowls readily eat both fish and flesh meat, and have no reluctance to feeding even on their own kind, picking much more faithfully than quadrupeds. Blood of any kind is esteemed by them a delicacy; and fish, even when salted, is devoured with a relish. They seem to be indifferent whether animal food is given to them in a cooked or raw state; though, if any preference can be detected, it is for the latter. They are sometimes so greedy that they will attack each other in order to taste the blood which flows from the wounds so inflicted; and it is quite common for them, in the moulting season, to gratify themselves by picking at the sprouting feathers on their own bodies and those of their companions. They appear to be partial to suet and fat; but they should not be allowed to devour these substances in large quantities, on account of their tendency to render them inconveniently fat.

It is highly advantageous to fowls to allow them a reasonable quantity of animal food for their diet, which should be fed to them in small pieces, both for safety and convenience. Bones and meat may be boiled; and the liquor, when mixed with bran or meal, is healthy, and not expensive.

INSECTS. Fowls have a decided liking to flies, beetles, grasshoppers, and crickets; and grubs, caterpillars, and maggots are held by them in equal esteem. It is difficult, however, to supply the poultry-yard with this species of food in sufficient quantity; but enough may be provided, probably, to serve as luxuries. Some recommend that pailfuls of blood should be thrown on dunghills, where fowls are allowed to run, for the purpose of enticing flies to deposit their eggs; which, when hatched, produce swarms of maggots for the fowls. With the same view, any sort of garbage or offal may be thrown out, if the dunghill is so situated—as it always should be—that its exhalations will not prove an annoyance.

LAYING.

The ordinary productiveness of a single individual of the family of domestic fowls is astonishing. While few hens are capable of hatching more than fifteen eggs, and are incapable usually of sitting more than twice in the year, frequent instances have occurred of hens laying three hundred eggs annually, while two hundred is the average number. Some hens are accustomed to lay at longer intervals than others. The habit of one variety is to lay once in three days only; others will lay every other day; and some produce an egg daily. The productiveness of hens depends, undoubtedly, upon circumstances, to a great degree. Climate has a great influence in this respect; and their lodging and food, as well as the care bestowed upon them, have more or less effect in promoting or obstructing their fecundity.

There seems to be, naturally, two periods of the year in which fowls lay—early in the spring, and in the summer; and this fact would seem to indicate that, if they were left to themselves, like wild birds, they would bring forth two broods in a year. The laying continues, with few interruptions, till the close of summer, when the natural process of moulting causes them to cease. This annual process commences about August, and continues through the three following months. The constitutional effect attending the beginning, continuance, and consequences of this period

ON THE WATCH.

—a very critical one in the case of all feathered animals—prevents them from laying, until its very close, when the entire coat of new feathers replaces the old, the washing of the nutritive juices, yielded by the blood for the express purpose of promoting this growth, is a great drain upon the system; and the constitutional forces, which would otherwise assist in forming the egg, are rendered inoperative. The approach of cold weather, also, at the close of the moulting period, contributes to the same result. As the season of moulting is every year later, the older the hen is, the later in the spring she will begin to lay. As pullets, on the contrary, do not moult the first year, they commence laying sooner than the elder hens; and it is possible, by judicious and careful management, so to arrange, in a collection of poultry tolerably numerous, as to have eggs throughout the year. It is a singular fact that pullets hatched very late in autumn, and therefore of stunted growth, will lay nearly as early as those hatched in spring. The checking of their growth seems to have a tendency to produce eggs; of course, very tiny ones at first.

When a hen is near to the time of laying, her comb and wattles change from their previous dull hue to a bright red, while the eye becomes more bright, the gait more spirited, and she occasionally cackles for three or four days. These signs rarely prove false; and when the time comes that she desires to lay, she appears very restless, going backward and forward, visiting every nook and corner, cackling meanwhile, as if displeased because she cannot suit herself with a convenient nest. Not having looked out for one previously, she rarely succeeds in pleasing herself till the moment comes when she can no longer tarry, when she is compelled to choose one of the boxes or baskets provided for this purpose in the poultry-house, where she settles herself in silence and lays.

In some instances, a hen will make choice of a particular nest in which to lay, and when she finds, upon desiring to lay, that this is pre-occupied by another hen, she will wait till it is vacated; but, in other cases, hens will go into any nest which they find, preferring, for the most part, those having the greatest number of eggs. The process of laying is, most probably, rather painful, though the hen does not indicate this by her cries; but the instant she has done she leaves the nest, and utters her joy by peculiarly loud notes, which are re-echoed by the cock, as well as by some of the other hens. Some hens, however, leave the nest in silence, after laying.

It seems ever to have been an object of great importance, in an economical point of view, to secure the laying of hens during those parts of the year when, if left to themselves, they are indisposed to deposit their eggs. For this purpose many methods have been devised, the most of which embrace an increase of rich and stimulating food. Some recommend shutting hens up in a warm place during winter, and giving them boiled potatoes, turnips, carrots, and parsnips. Others assign as the reason for their not laying in winter, in some climates, that the earth is covered with snow, so that they can find no ground, or other calcareous matter, to form the shells; and advise, therefore, that bones of meat or poultry should be pounded and given to them, either mixed with their food, or by itself, which they will greedily eat. Upon the whole, it would seem that the most feasible means of obtaining fresh eggs during the winter is to have young hens—pullets hatched only the previous spring being the best—to use extreme liberality in feeding, and to cautiously abstain from over-stocking the poultry-yard.

As serviceable *food* to increase laying, scraps of animal food, given two or three times a week, answer admirably; the best mode of doing so is throwing down a bullock's liver, leaving it with them, and permitting them to pick at it will; this is better raw than boiled. Lights, or guts, or any other animal refuse, will be found to answer the same purpose; but these substances require, or, at all events, are better for, boiling. Cayenne pepper—in fact all descriptions of pepper, but especially cayenne pepper in pods—is a favorite food with fowls; and, being a powerful stimulant, it promotes laying.

An abundant supply of lime, in some form, should not be omitted; either chopped bones, old mortar, or a lump of chalky marl. The shell of every egg used in the house should be roughly crushed and thrown down to the hens, which will greedily eat them. A green, living turf will be of service, both for its grass and the insects it may contain. A dusting-place, wherein to get rid of vermin, is indispensable. A daily hot meal of potatoes, boiled as carefully as for the family table, then chopped, and sprinkled or mixed with bran, will be comfortable and stimulating. After every meal of the household, the bones and other scraps should be collected and thrown out.

As to *the number of eggs*, the varieties which possess the greatest fecundity are the Shanghaes, Guelderlands, Dorkings, Polish, and Spanish. The Poland and Spanish lay the largest eggs; the Dorkings, eggs of good size; while the Game and the smaller kinds produce only small eggs. Those eggs which have the brightest yolks are the finest flavored; and this is usually the case with the smaller kinds. The large eggs of the larger varieties often have yolks of a pale color, and are inferior in flavor.

PRESERVATION OF EGGS.

Eggs, after being laid, lose daily, by transpiration, a portion of the matter which they contain, notwithstanding the compact texture of their shell, and of the close tissue of the flexible membranes lining the shell, and enveloping the white. When an egg is fresh, it is full, without any vacancy; and this is a matter of common observation, whether it be broken raw, or when it is either soft or hard-boiled. In all stale eggs, on the contrary, there is uniformly more or less vacancy, proportioned to the loss they have sustained by transpiration; hence, in order to judge of the freshness of an egg, it is usual to hold it up to the light, when the transparency of the shell makes it appear whether or not there is any vacancy in the upper portion, as well as whether the yolk and white are mingled and muddy, by the rotting and bursting of their enveloping membranes.

The transpiration of eggs, besides, is proportional to the temperature in which they are placed, cold retarding and heat promoting the process; hence, by keeping fresh-lain eggs in a cool cellar, or, better still, in an ice-house, they will transpire less, and be preserved for a longer period sound, than if they are kept in a warm place, or exposed to the sun's light, which has also a good effect in promoting the exhalation of moisture. As, therefore, fermentation and putridity can only take place by communication with the air at a moderate temperature, such connection must be excluded by closing the pores of the shell.

It is an indispensable condition of the material used for this purpose, that it shall be incapable of being dissolved by the moisture transpired from the interior. Spirits of wine varnish, made with lac, answers the requirement; this is not very expensive, but is rather an uncommon article in country places, where eggs are most abundantly produced.

A better material is a mixture of mutton and beef suet, which should be melted together over a slow fire, and strained through a linen cloth into an earthen pan. The chief advantage in the use of this is, that the eggs rubbed over with it will boil as quickly as if nothing had been done to them, the fat melting off as soon as they touch the water. The transpiration is as effectually stopped by the thinnest layer of fat as by a thick coating, provided that no sensible vestige be left on the surface of the shell. All sorts of fat, grease, or oil are well adapted to this purpose; by means of butter, hog's lard, olive oil, and similar substances, eggs may be preserved for nine months as fresh as the day upon which they were laid.

Another method is, to dip each egg into melted pork-lard, rubbing it into the shell with the finger, and pack them in old fig-drums, or butter firkins, setting every egg upright, with the small end downward. Or, the eggs may be packed in the same way in an upright earthen pan; then cut some rough sheep's tallow, procured the same day that the animal is killed, into small pieces, and melt it down; strain it from the scraps, and pour it while warm, not hot, over the eggs in the jar till they are completely covered. When all is cold and firm, set the vessel in a cool, dry place till the contents are wanted.

Eggs will also keep well when preserved in salt, by arranging them in a barrel, first a layer of salt, then a layer of eggs, alternately. This can, however, also act mechanically, like bran or saw-dust, so long as the salt continues dry; for, in that case, the chlorine, which is the antiseptic principle of the salt, is not evolved. When the salt, however, becomes damp, its preservative principle will be brought into action, and may penetrate through the pores of the shell.

Immersing eggs in vitriol, or sulphuric acid, is likewise a very effectual means of preserving them; the sulphuric acid acts chemically upon the carbonate of lime in the shell, by setting free the carbonic acid gas, while it unites with the lime, and forms sulphate of lime, or plaster of Paris. Another method is, to mix together a bushel of quick-lime, two pounds of salt, and eight ounces of cream of tartar, adding a sufficient quantity of water, so that eggs may be plunged into the paint. When a paste is made of this consistence, the eggs are put into it, and may be kept fresh, it is said, for two years.

Another method of preserving eggs a long while fresh, depends upon a very different principle. Eggs that have not been rendered reproductive by the cock have been found to continue very uncorrupted. In order, therefore, to have eggs keep fresh from spring to the middle or even to the end of winter, it is only necessary to deprive the hens of all communication with the cocks, for at least a month before the eggs are put away.

It ought not to be overlooked, in this connection, that eggs not only spoil by the transpiration of their moisture and the putrid fermentation of their contents, in consequence of air penetrating through the pores of the shell, but also by being moved about and jostled, when carried to a distance by sea or land. Any kind of rough motion, indeed, ruptures the membranes which keep the white, the yolk, and the germ of the chicken in their appropriate places; and, upon these being mixed, putrefaction is promoted.

CHOICE OF EGGS FOR SETTING.

Eggs for hatching should be as fresh as possible; if laid the very same day, so much the better. This is not always possible when a particular stock is required; but, if a numerous and healthy brood is all that is wanted, the most recent eggs should be selected. Eggs may be kept for this purpose in either of the ways first mentioned; or they may be placed on their points in a box, in a cool, dry place; the temperature about sixty or sixty-five, Fahrenheit; the bottom of the box should be covered with a layer of wheat bran, then a layer of eggs put in, and covered with bran; and so on, alternating. In this mode, evaporation is prevented, and the eggs are almost as certain to hatch out, at the end of six weeks, or even two months, as when they were laid.

It is difficult to fix the exact term during which the vitality of an egg remains unextinguished; as it, unquestionably, varies from the very first, according to the vigor of the parents of the inclosed germ, and fades away gradually till the final moment of non-existence. The chickens in stale eggs have not sufficient strength to extricate themselves from the shell; if assisted, the yolk is found to be partially absorbed into the abdomen, or not at all; they are too faint to stand; the muscles of the neck are unable to lift their heads, much less to peck; and although they may sometimes be saved by extreme care, their usual fate is to be trampled to death by the mother, if they do not expire almost as soon as they begin to draw their breath. Thick-shelled eggs, like those of geese, Guinea fowls, etc., will retain life longer than thin-shelled ones, as those of hens and ducks. When choice eggs are expected to be laid, it is more prudent to have the hen which is to sit upon them wait for them, than to keep other eggs waiting for her. A good sitter may be amused for two or three weeks with a few addle-eggs, and so be ready to take charge of those of value immediately upon their arrival.

As to the choice of eggs for hatching, such should be taken, of course, as are believed to have been rendered productive. Those of medium size—the average size that the hen lays—are most apt to fulfil this requirement. A very fair judgment may be formed of eggs from their specific gravity; such as do not sink to the bottom in a bowl of tepid water should be rejected.

The old-time notion, that small, round eggs produce females, and long, pointed ones males—originally applied, by the ancients, to eating rather than hatching purposes—may be considered exploded. The hen that lays one round egg, continues to lay all her eggs round; and the hen that lays one oblong, lays all oblong. According to this theory, then, one hen would be the perpetual mother of cocks, and another the perpetual producer of pullets; which is absurd, as daily experience proves.

The same fate has been meted out to that other venerable test of sex, the position of the air-bag at the blunt end of the shell. "If the vacancy is a little on one side, it will produce a hen; if it is exactly in the centre, a cock." Upon this assumption, the cock should be a very rare bird; since there are very few eggs indeed in which the air bottle is exactly concentric with the axis of the egg. In many breeds, on the contrary, the cockerels bear a proportion of at least one-third, and sometimes two-thirds, especially in those hatched during winter, or in unfavorable seasons; the immediate cause, doubtless, being that the eggs producing a more robust sex possess a stronger vitality.

Nor are these two alleged tests—the shape of the egg, and the position of the air-tube—consistent with each other; for, if the round egg produces a pullet, and an egg with the air-bag a little on one side does the same, then all round eggs should have the air bag in that position, or one test contradicts the other; and the same argument applies to the long or oval egg. The examination of a few eggs by the light of a candle will satisfy any one that the position of the air-bag differs as much in a long egg as it does in a round.

There are, indeed, no known means of determining beforehand the sex of fowl; except, perhaps, that cocks may be more likely to issue from large eggs, and hens from small ones. As, however, the egg of each hen may be recognized, the means are accessible of propagating from those parents whose race it is judged most desirable to continue.

INCUBATION.

The hen manifests the desire of incubation in a manner different from that of any other known bird. Nature having been sufficiently tasked in one direction, she becomes feverish, and loses flesh; her comb is livid; her eyes are dull; she bristles her feathers to intimidate an imaginary enemy; and, as if her chickens were already around her, utters the maternal "cluck."

When the determination to sit becomes fixed—it is not necessary to immediately gratify the first faint inclinations—the nest which she has selected should be well cleaned, and filled with fresh straw. The number of eggs to be allowed will depend upon the season, and upon the size of egg and hen. The wisest plan is not to be too greedy; the number of chickens hatched is often in inverse proportion to the number of eggs set—five have only been obtained from sixteen. An odd number is, however, to be preferred, as being better adapted to covering in the nest. Hens will, in general, well cover from eleven to thirteen eggs laid by themselves. A bantam may be trusted with about half a dozen eggs of a large breed, such as the Spanish. A hen of the largest size as a Dorking, will successfully hatch, at the most, five goose-eggs.

When hens are determined to sit at seasons of the year at which there is little chance of bringing up chickens, the eggs of ducks or geese may be furnished her; the young may be reared, with a little painstaking, at any time of the year. The autumnal laying of the China and of the common goose is very valuable for this purpose. Turkey-hens frequently have this fit of unseasonable incubation.

Where, however, it is inconvenient to gratify the desire, one or two doses of jalap will often entirely remove it; and fowls often lay in three weeks afterward. Some place the would-be sitter in an aviary, for four or five days at most, and feed her but sparingly; from the commencement of her confinement, she will gradually leave off clucking, and when this has ceased, she may be again set free, without manifesting the least desire to take to the nest again, and in a short time the hen will commence laying with renewed vigor. The barbarous measures sometimes resorted to should be frowned upon by every person with humane feelings.

Three weeks is the period of incubation; though chickens are sometimes excluded on the eighteenth day. When the hen does not sit close for the first day or two, or in early spring, it will occasionally be some hours longer; when the hen is assiduous, and the weather hot, the time will be a trifle shorter. Chickens have been known to come out as late as the twenty-seventh day.

It may not be uninteresting to note the changes which the egg passes through in hatching. In *twelve hours*, traces of the head and body of the chicken may be discerned; at the end of the *second day*, it assumes the form of a horse-shoe, but no red blood as yet is seen; at the *fiftieth hour*, two vesicles of blood, the rudiments of the heart, may be distinguished, one resembling a noose folded down on itself, and pulsating distinctly; at the end of *seventy hours*, the wings may be seen, and, in the head, the brain and the bill, in the form of bubbles; toward the end of the *fourth day*, the heart is more completely formed; and on the *fifth day*, the liver is discernible; at the end of *one hundred and thirty hours*, the first voluntary motions may be observed; in *seven hours* more, the lungs and stomach appear; and, in *four hours* after this, the intestines, the loins, and the upper jaw. At the end of the *one hundred and forty-fourth hour*, two drops of blood are observable in the heart, which is also further developed; and, on the *seventh day*, the brain exhibits some consistence. At the *one hundred and ninetieth hour*, the bill opens, and the muscular flesh appears on the breast; in *four hours* more, the breast bone is seen; and, in *six hours* afterward, the ribs may be observed forming from the back. At the expiration of *two hundred and thirty-six hours*, the bill assumes a green color, and, if the chicken be taken out of the egg, it will visibly move. At *two hundred and sixty-four hours*, the eyes appear; at *two hundred and eighty eight hours*, the ribs are perfect; and at *three hundred and thirty-one hours*, the spleen approaches near to the stomach, and the lungs to the chest; at the end of *three hundred and fifty-five hours*, the bill frequently opens and shuts. At the end of the *eighteenth day*, the first cry of the chicken is heard; and it gradually acquires more strength, till it is enabled to release itself from confinement.

After the hen has set a week, the fertility of the eggs may be satisfactorily ascertained by taking a thin board with a small orifice in it, placing a candle at the back, and holding up each egg to the points of light. The barren eggs may then be removed, and used, hard-boiled, for young chickens. Some reserve this for the eleventh or twelfth day.

About the *twenty-first day*, the chicken is excluded from the egg; for the purpose of breaking the shell of which it is furnished with a horny-pointed scale, greatly harder than the bill itself, at the upper tip of the bill—a scale which falls off, or becomes absorbed, after the chicken is two or three days old. The chicken is rolled up in the egg in the form of a ball, with its forepart toward the highest end, and its beak uppermost, the hard scale nearly touching the shell.

The first few strokes of the chicken's beak produce a small crack, rather nearer the larger than the smaller end of the egg, and the egg is said to be *chipped*. From the first crack, the chicken turns gradually round, from left to right, chipping the shell as it turns, in a circular manner, never obliquely. All do not succeed in producing the result in the same time; some being able to complete the work within an hour, and others taking two or three hours, while half a day is most usually employed, and some require twenty-four hours, or more, but rarely two days. Some have greater obstacles to overcome than others, all shells not being alike in thickness and hardness.

When chickens do not effect their escape easily, some little assistance is needed; but the difficulty is to know when to give it, as a rash attempt to help them, by breaking the shell, particularly in a downward direction toward the smaller end, is often followed by a loss of blood, which can ill be spared. It is better not to interfere, until it is apparent that a part of the brood have been hatched for some time, say twelve hours, and that the rest cannot succeed in making their appearance. It will then generally be found that the whole fluid contents of the egg, yolk and all, are taken up into the body of the chicken, and that weakness alone has prevented its forcing itself out. The causes of such weakness are various; sometimes, insufficient warmth, from the hen having set on too many eggs; sometimes the original feebleness of the vital spark; but, most frequently, the staleness of the eggs employed for incubation.

The chances of rearing such chickens are small; but, if they survive the first twenty-four hours, they may be considered as safe. The only thing to be done is to take them from the hen till she is settled at night, keeping them in the meanwhile as snug and warm as possible. If a gentle hand can persuade a crust of bread down their throats, it will do no harm; but all rough and clumsy manipulation will utterly defeat the end in view. Animal heat will be their greatest restorative. At night, they should be quietly slipped under their mother; the next morning will disclose the sequel.

The period of incubation in the *Guinea fowl* is twenty-eight days, or one month; in the *pea fowl*, from twenty-seven to twenty-nine days; in *turkeys*, a month; in *ducks*, thirty or thirty-one days; and in *geese*, from twenty-seven to thirty days.

INCUBATION OF TURKEYS. When the turkey hen has once selected a spot for her nest, she will continue to lay there till the time for incubation; so that the egg may be brought home from day to day, there being no need of a nest-egg, as with the domestic fowl. She will lay from fifteen to twenty eggs, more or less. If there are any dead leaves or dry grass at hand, she will cover her eggs with these; but if not, she will take no trouble to collect them from a distance.

Her determination to sit will be known by her constantly remaining on the nest, though it is empty; and, as it is seldom in a position sufficiently secure against the weather or pilferers, a nest should be prepared for her, by placing some straw, with her eggs, on the floor of a convenient out-building. She should then be brought home, and gently and kindly placed upon it. With the smallest varieties, thirteen eggs will suffice; a large hen might cover more. At the end of a week, it is usual to add some fowls' eggs; the activity of the chickens excites some emulation in the larger brethren, and the eggs take up but little room in the nest.

Some believe it necessary to turn the eggs once a day; but the hen herself does that many times daily. If the eggs are marked, and their position noticed when she leaves the nest, they will never be found in the same order. In about four weeks, the young will be hatched.

INCUBATION OF GEESE. Geese breed in general only once a year; but, if well kept, they sometimes hatch twice a season. During the sitting, in sections where the most attention is paid to breeding them, each bird has a space allotted to it, in rows of wicker-pens, placed one above another, and the person in charge of them drives the whole flock to water three times a day, and, bringing them back to their habitations, places each bird in its own nest.

The most successful breeders of *Bremen geese* adopt the

following method: The birds are, in the first place, carefully and properly fed; the eggs are removed every day in the gentlest manner from the nest, and placed in a basket of cotton kept in a moderate temperature, and free from damp. When all the geese begin to sit steadily, each is furnished with a nest composed of chopped straw; and care is taken that it is sufficiently capacious.

Not more than one of the geese is allowed to leave the eggs at a time. As soon as one leaves, she makes a cackling noise, which is the signal for the attendant to shut up the boxes in which the others are sitting. These are made somewhat like a dog-kennel, with a roof pitched both ways; and are thirty inches long, by twenty-four wide, and twenty-four high; the door is in the end, and is covered by a sliding panel, which moves upward, when egress or ingress is sought, and may be shut down at pleasure. The goose, upon returning, finds only her own box open. When she re-enters her box, the whole of the doors are again opened, and the same rule observed throughout the period of hatching. In this way, each goose is kept to its own nest.

REARING OF THE YOUNG.

For about twenty-four hours after birth, the chickens can not only do well enough without any extraneous nourishment, but will be far more likely to thrive subsequently, if let alone, than if crammed or incited to eat prematurely. More chickens are destroyed by over-feeding than are lost by the want of it. It is, however, well to turn them in among other chickens that already feed themselves; they will, in such cases, generally follow the example of the rest, and pick away at whatever is around.

A roomy, boarded coop, in a dry, sunny spot, is the best position for them during the first month; after which it may be left open during the day, for the hen to retire to when she pleases. In quiet, grassy places, it is scarcely necessary to coop the hen at all. As to food, they may have every thing which is not absolutely poisonous; though if wet food is given, the chicken is thus obliged to take water, whether it requires it or not, in order to get a sufficient supply of solid food, and diseased bowels will be likely to follow; whereas, if the food is dry, they can supply themselves with food and water according to their pleasure. If Indian meal is well boiled, and fed not too moist, it will answer a very good purpose, particularly after they are eight or ten days old. Pure water must be placed near them in such a manner as to enable them to drink without getting into the water, which, by wetting their feathers, benumbs and injures them. Meat and insect diet are almost necessary; but, whatever the food, the meals must be given at short intervals; as much as they can swallow, and as often as they can eat. With all their industry, they are only half-clad till flesh and bone stop growing for a while, and allow down and feathers to overtake them.

MARQUEE, OR TENT-SHADED COOPS.

Chickens should not be let out of their coops too early in the morning, or whilst the dew is on the ground; still less should they be suffered to range over the wet grass, which is a common cause of disease and death. They should also be guarded against sudden unfavorable changes of the weather, more particularly if attended with rain. Nearly all the diseases of gallinaceous fowls arise from cold moisture.

The period at which they are left to shift for themselves depends upon the disposition of the hen. Some will continue their attentions to their chickens till they are nearly full-grown, while others will cast them off much earlier. In the latter case, an eye should be kept upon them for a few days; for chickens in this half-grown state are much more liable to disease than when they were apparently tender little weaklings, crowded under their mother's wings. They should be kept in a dry, warm, place; dryness is especially necessary.

If the chickens feather rapidly when very young—as is the case with the Golden Pheasant, Black Poland, Guelderland, and some others—they are always weakly, however healthy in other respects, from the fact that their food goes to sustain their feathers rather than their bodies; and they frequently languish and die, from this circumstance alone. If, on the other hand, they feather slowly, as do the Cochin Chinas, Shanghaes, and others, the food in early life goes to nourish and sustain their bodies until they become more vigorous, and old enough to sustain the shock of feathering without detriment. Pure tan-colored Dorkings are more easily raised than others of the race, because they feather more slowly.

Chickens which feather rapidly must be kept perfectly dry and warm, or they will die; while naked chickens, as they are termed, or those which feather at a more advanced age, and very slowly, seldom suffer from the cold, from the fact that their down is very warm, and their blood is hotter, and circulates more rapidly; since their food principally goes to blood, and flesh, and bones, and not to feathers.

Rearing of Guinea fowls. For the young of these, ants' eggs, so called, hard-boiled eggs chopped fine, small worms, maggots, bread-crumbs, chopped meat, or suet—whatever, in short, is most nutritious, is the most appropriate food. This need not be offered to them in large quantities, as it would only be devoured by the mother Bantam as soon as she saw that they had for the time satisfied their appetites, or it would be stolen by other birds; but it should frequently be administered to them in small supplies. Feeding them three, four, or five times a day, is not often enough; every half hour during daylight they should be tempted to fill their craws, which are soon emptied again by an extraordinary power and quickness of digestion.

The newly-hatched Guinea fowl is a tiny creature, and its growth is, consequently, very rapid, requiring incessant supplies. A check once received can never be recovered. They do not, in such cases, mope and pine for a day or two, like young turkeys under similar circumstances, and then die; but, in half an hour after being in apparent health, they fall on their backs, give a convulsive kick or two, and fall victims, in fact, to starvation. The demands of Nature for the growth of bone, muscle, and particularly of feathers, are so great, that no subsequent abundant supply of food can compensate for a fast of a couple of hours. The feathers still go on growing in geometrical progression, and drawing the sources of vitality still faster than they can be supplied, till the bird faints and expires from inanition.

A dry, sunny corner in the garden will be the best place to

coop them with their bantam hen. As they increase in strength, they will do no harm, but much good, by devouring worms, grubs, caterpillars, maggots, and all sorts of insects. By the time their bodies are little longer than those of sparrows, they will be able to fly with some degree of strength; other additions to their complete stature are successively and less immediately developed, the spurs, comb, and ornamental plumage not appearing till a subsequent period.

When they are about the size of thrushes, or a little larger, unless the summer be very fine, the bantam may be allowed to range loose in the orchard and shrubbery, and no longer permitted to enter the garden. The young must, however, still receive a bountiful and frequent supply of food; they are not to be considered safe till the horn on their head is fairly grown. Oatmeal is a great treat; cooked potatoes, boiled rice, or any thing, in short, that is eatable, may be thrown to them; they will pick the bones left after dinner with evident satisfaction. The tamer they can be made, the less troublesome will they be when grown; the more kindly they are treated, the fatter will they be for food, and the better price will they bring in market.

For rearing the young of the *pea fowl*, the same directions will be found useful, and should be carried out in practice.

Rearing of turkeys. Much quackery has been recommended in the treatment of young turkeys. Nothing, however, should be given to them, nothing done for them; they should remain in the nest, under the shelter of their mother's wings, for at least eight or ten hours; if hatched in the afternoon, till the following morning. The hen should then be placed on the grass, in the sun, under a roomy coop. If the weather is fine, she may be stationed at any point desired, by a long piece of flannel-list tied round one leg, and fastened to a stump or stone.

At first, a few crumbs of bread should be offered; for some hours, the little ones will be in no hurry to eat; but, when they do commence, they should be supplied constantly and abundantly with chopped egg, shreds of meat and fat, curd, boiled rice mixed with cress, lettuce, and the green of onions; melted mutton-suet poured over barley-meal, and cut up when cold, as also bullock's liver boiled and minced, are excellent things. Young turkeys do not like to have their food minced much smaller than they can swallow it, preferring to make a meal at three or four mouthfuls, rather than to trouble themselves with the incessant pecking and scratching in which chickens so much delight. Pepper will be found particularly useful in feeding them; as, indeed, all stimulating vegetables, such as horse-radish, and the like.

Young turkeys are sometimes attacked by *fasciolæ*, or worms in the trachea; but not so often as chickens. Cramp is the most fatal to them, particularly in bad weather. A few pieces of board laid under and about the coop are useful; sometimes rubbing the leg with spirit will bring back the circulation.

The time when the hen may be allowed full liberty with her brood depends most upon the season, the situation, etc. Some think that if the young are thriving, the sooner the old ones are out with them the better, after the first ten days or so. A safer rule may be fixed at the season, called "shooting the red," when young turkeys approach the size of a partridge, or before the granular, fleshy excrescences on the head and neck begin to appear; soon after, the whole plumage, particularly the tail-feathers, shoot into rapid growth, and liberal nourishment is imperatively required If let loose at this time, they will obtain much foraging, and still be thankful for all that is given to them. Caraway-seeds, as a tonic, are beneficial, if added to plenty of barley, boiled potatoes, chopped vegetables, and refuse meat. At this time the turkeys, naturally enough, begin to be troublesome and voracious; they have to grow from the size of a lark to twelve or fourteen pounds, in eight or nine months. One great merit in old birds is, that in situations where nuts, acorns, and mast are to be had, they will lead off their brood to these, and all of them will abstain, comparatively, from ravaging other crops.

Rearing of ducklings. The best mode of rearing the young of ducks depends very much upon the situation in

DUCK-POND AND HOUSES.

which they are hatched. It is customary to dip their feet in water as soon as they are hatched, and then to clip the down on their tails close with a pair of scissors, to prevent their becoming drabbled and water-logged; and before their introduction to the pond, which should not be until a day or two after hatching, it is thought advisable by many to let them have a private swim or two in a small pan of water, that they may try their strength and practice their webbed feet before venturing upon a larger space.

For the first month, the confinement of the mother under a coop is better than too much liberty. Their first food may be boiled eggs, nettles, and a little barley; all kinds of sapped food, cornmeal and water mixed thin, worms, etc., suit them; they will also greedily eat cabbages or other greens, mixed with boiled bran; and this mess, with the addition of pepper, forms a valuable dietetic. In a few days, they require no care, being perfectly able to shift for themselves; but at any age they are the most helpless of the inhabitants of the poultry-yard, having no weapons with which to defend themselves from vermin, or animals of prey, and their awkward, waddling gait precluding their seeking safety in flight. The old duck is not so brave in defence of her brood as the hen; but she will, nevertheless, display at times much spirit. The young seldom die of any disease, and with proper precaution there will be no trouble in raising almost as many ducklings as are hatched. They come early to maturity, being nearly full-grown and in fine eating order at three months old; far excelling, in this respect, all other poultry, except geese.

None are more successful in rearing ducklings than those who keep them, for the first period of their existence, in pens two or three yards square, and cram them night and morning with long, dried pellets of flour and water, or egg and flour, until they are judged old enough to be turned out with their mother to forage for themselves. They are cheerful, harmless, good-natured, cleanly creatures, carefully washing themselves, and arranging their dress, before commencing their meals; and the healthy heartiness of their appetite is amusing, rather than disgusting.

REARING OF GOSLINGS. For the first three or four days, goslings must be kept warm and dry, and fed on barley-meal, or oatmeal, mixed with milk, if easily procurable; if not, with water. They will begin to grow in about a week. For a week or two, they should not be turned out until late in the morning, and should always be taken in early in the evening. Their great enemy is the cramp, which can be kept off by making them sleep on dry straw. A little boiled rice, daily, assists their growth; with corn, of course, as soon as they can eat it. When goslings are first allowed to go at large with their mother, every plant of hemlock which grows within their range should be pulled up, as they are very apt to eat it, and it generally proves fatal. Nightshade is equally pernicious to them; and they have been known to be poisoned by eating sprigs of yew-tree.

The young of *Bremen geese*, when first hatched, are of a very delicate and tender constitution. It is best to let them remain in the breeding-box in which they are hatched for twenty-four hours after they leave the shell. This should, however, be regulated by the weather; since, if it is fair and warm, they may be let out an hour or two in the middle of the day, when they will wet their little bills and nibble at the grass. They ought not to be out in the rain at any time during the first month; and both geese and goslings should be shut up in the boxes at night, during the same period, as a protection against rats and vermin. A very shallow pool, dug in the yard, with a bucket or two of water thrown into it, to suit the temporary purpose of bathing, is sufficient during that period. If well fed on grain from the time they are hatched, twenty-five pounds weight can be secured, at seven or eight months old. By feeding them till four days old, and then literally turning them out to grass, an average weight of from seventeen to eighteen pounds each has been attained, at that age after the feathers are cleanly picked off.

CAPONIZING.

Capons have ever been esteemed among the greatest delicacies of the table; and are made by the extirpation of the reproductive organs in male fowls. If a cock, when young, is emasculated, a remarkable change takes place in him. His natural fierceness is calmed; he becomes placid and peaceful; his pugnacity has deserted him; he no longer seeks the company of the hens; he loses his previous strong, shrill voice; he grows to a far larger size than he would otherwise have done, having nothing to interfere with the main business of his life—to eat, drink, sleep, and get fat as speedily as possible; his flesh is peculiarly white, firm and succulent; and even the fat is perfectly destitute of rankness. The capon may, also, by a little management be converted into an admirable nurse. Some assert that caponized cocks are never afterward subject to the natural process of moulting; but this is denied by others.

The art has been practised from the earliest antiquity, in Greece, India, and China, for the purpose of improving the flesh of birds for the table, in tenderness, juiciness, and flavor. It is extensively performed in the great poultry-breeding districts of England; but in this country it is by no means so generally practised as would naturally be expected.

The instruments most approved by skilful operators consist of two five or seven-pound weights for confining the fowl; a scalpel, for cutting open the thin skin enveloping the testicles; a silver retractor, for stretching open the wound sufficiently wide for operating within; a pair of spring forceps—with a sharp, cutting edge, resembling that of a chisel, having a level half an inch in its greatest width—for making the incision, and securing the thin membrane; a spoon-shaped instrument, with a sharp hook at one end, for pushing and removing the testicles, adjusting the loop, and assisting in tearing open the tender covering; and a double silver canula, for containing the two ends of horse-hair, or fibre, constituting the loop. The expense of these instruments is in the neighborhood of six dollars. A cheap penknife may be used instead of the scalpel; and the other instruments may be obtained of a cheaper construction—the whole not costing more than half the above-named amount.

The cockerel intended for capons should be of the largest breeds, as the Dorking, Cochin China, or the Great Malay. They may be operated upon at any time after they are a month old; the age of from two to three months is considered preferable. If possible, it should be done before July; as capons made later never prove so fine.

The fowl should be confined to a table or board, by laying him with the left side downward, the wings drawn behind the rump, the legs extended backward, with the upper one farthest drawn out, and the head and neck left perfectly free. The feathers are next to be plucked from the right side, near the hip-joint, on a line with, and between the joint of the shoulder. The space uncovered may be from an inch to an inch and a half in diameter, according to the size of the bird. After drawing off the skin from the part, backward—so that, when left to itself after the operation is completed, it will cover the wound in the flesh—make an incision with the bevel-edged knife, at the end of the forceps, between the last two ribs, commencing about an inch from the back-bone, and extending it obliquely downward, from an inch to an inch and a half, cutting just deep enough to separate the ribs, taking due care not to wound the intestines.

Next, adjust and apply the retractor by means of the small thumb-screw, and stretch the wound sufficiently wide apart to afford room for an examination of the organs to be removed. Then, with the scalpel, or a sharp penknife, carefully cut open the skin, or membrane, covering the intestines, which, if not sufficiently drawn up, in consequence of the previous confinement, may be pushed forward toward the breast-bone, by means of the bowl of the spoon-shaped instrument, or—what would answer equally well—with the handle of a tea-spoon.

As the testicles are exposed to view, they will be found connected with the back and sides by a thin membrane, or skin, passing over them. This covering must then be seized with the forceps, and torn open with the sharp-pointed hook at the small end of the spoon-shaped instrument; after which the bowl of the spoon must be introduced, with the left hand, under the lower or left testicle, which is, generally, a little nearer to the rump than the right one. Then take the double canula, adjust the hair-loop, and, with the right hand, pass the loop over the small hooked end of the spoon, running it down under the bowl of the spoon containing the testicle, so as to bring the loop to act upon the parts which connect the testicle to the back. By drawing the ends of the hair-loop backward and forward, and at the same time pushing the lower end of the tube, or canula, toward the rump of the fowl, the cord or fastening of the testicle is severed.

A similar process is then to be repeated with the uppermost or right testicle; after which, any remains of the testicles, together with the blood at or around the bottom of the wound, must be scooped out with the bowl of the spoon. The left testicle is first cut out, in order to prevent the blood which may issue from covering the one remaining, and so rendering it more difficult to be seen. The operation, if skilfully done, occupies but a few moments; when the skin of the fowl should be drawn over the wound with the retractor, and the wound covered with the feathers that were plucked off at the commencement.

In some fowls, the fore part of the thigh covers the two hindmost ribs; in which case, care must be taken to draw the fleshy part of the thigh well back, to prevent it from being cut; since, otherwise, the operation might lame the fowl, or even cause its death.

For loops, nothing answers better than the fibre of a cocoanut husk, which is rough, and readily separates the testicles by sawing. The next best substance is the hair of a horse's mane or tail.

After the operation, the bird may be placed in a warm house, where there are no perches; since if such appliances are present, the newly-made capon will very probably injure himself in his attempts to perch. For about a week, the food should be soft, meal porridge, and that in small quantities, alternated with bread steeped in milk; he may be given as much pure water as he will drink, it being best to use it in a tepid state, or at least with the chill taken off. At the end of a week, or ten days, at most, the fowl, if previously of a sound, vigorous constitution, will be all right, and may be turned out with the others.

The usual method, in France, of making *poulardes*, or hen-capons, as they are sometimes improperly designated, is to extirpate the egg-cluster, or *ovarium*, in the same manner as the testicles are extracted from the cockerel; but it is quite sufficient merely to cut across the oviduct, or egg-tube, with a sharp knife. Otherwise, they may be treated in the same manner as the capons. Capons are fattened in precisely the same manner as other fowls.

FATTENING AND SLAUGHTERING.

Fat is not a necessary part of any animal body, being the form which superabundant nourishment assumes, which would,

A BAD STYLE OF SLAUGHTERING.

if needed, be converted into muscles and other solids. It is contained in certain membranous receptacles provided for it, distributed over the body, and it is turned to use whenever the supply of nourishment is defective, which should be provided by the stomach, and other great organs. In such emergencies it is taken up, in the animal economy, by the absorbents; if the latter, from any cause, act feebly, the health suffers. When, however, nourishment is taken into the system in greater quantities than is necessary for ordinary purposes, the absorbent vessels take it up; and the fat thus made is generally healthy, provided there is a good digestion.

A common method of fattening fowl is to give them the run of a farm-yard, where they thrive upon the offal of the stable and other refuse, with perhaps some small regular daily feeds; but at threshing-time, they become fat, and are styled *barn-door fowls*, probably the most delicate and high-flavored of all, both from their full allowance of the finest grain, and the constant health in which they are kept, by living in the natural state, and having the full enjoyment of air and exercise; or, they are confined in coops during a certain number of weeks, those fowls which are soonest ready being taken as wanted.

Fowls may also be fattened to the highest pitch, and yet preserved in a healthy state—their flesh being equal in quality to that of the barn-door fowl—when confined in feeding-houses. These should be at once warm and airy, with earth floors, well-raised, and sufficiently capacious to accommodate well the number desired. The floor may be slightly littered down, the litter being often changed; and the greatest cleanliness should be observed. Sandy gravel should be placed in several different layers, and often changed. A sufficient number of troughs, for both water and food, should be placed around, that the fowls may feed with as little interruption as possible from each other; and perches in the same proportion should be furnished for those which are inclined to avail themselves of them; though the number will be few, after they have begun to fatten. This arrangement, however, assists in keeping them quiet and contented until that period. Insects and animal food forming a part of the natural diet of poultry, they are medicinal to them in a weakly state, and the want of such food may sometimes impede their thriving.

The least nutritious articles of food, so far as it can be done conveniently, should be fed out first; afterward, those that are more nutritive. Fattening fowls should be kept quiet, and suffered to take no more exercise than is necessary for their health; since more exercise than this calls for an expenditure of food which does not avail any thing in the process of fattening. They should be fed regularly with suitable food, and that properly prepared; and as much should be given them as they are able to convert into flesh and fat, without waste. The larger the quantity of food which a fattening animal can be made to consume daily, with a good appetite, or which it can digest thoroughly, the greater will be the amount of flesh and fat gained, in proportion to the whole quantity of food consumed.

Substances in which the nutriment is much concentrated should be fed with care. There is danger, especially when the bird is first put to feed, that more may be eaten at once than the digestive organs can manage. Meal of Indian corn is highly nutritive; and, when properly fed, causes fowls to fatten faster than almost any other food. They will not, however, bear to be kept exclusively on this article for a great length of time. Meal made from the heaviest varieties of corn, especially that made from the hard, flinty kinds grown in the

Northern and Eastern States, is quite too strong for fowls to be full-fed upon. Attention should also be paid to the bulk of the food given; since sufficient bulk is necessary to effect a proper distending of the stomach, as a necessary condition of healthy digestion.

One simple mode of fattening, which is adopted by many, is the following: Shut the fowls up where they can get no gravel; keep corn by them all the time, and also give them dough enough once a day; for drink, give them skimmed milk; with this feed, they will fatten in ten days; if kept longer, they should have some gravel, or they will fall away.

Oats ground into meal, and mixed with a little molasses and water, barley-meal with sweet milk, and boiled oats, mixed with meat, are all excellent for fattening poultry—reference being had to time, expense, and quality of flesh.

In *fattening ducks*, it must be remembered that their flesh will be found to partake, to a great extent, of the flavor of the food on which they have been fattened; and as they are naturally quite indiscriminate feeders, care should be taken, for at least a week or so before killing, to confine them to select food. Boiled potatoes are very good feeding, and are still better if a little grain is mixed with them; Indian meal is both economical and nutritive, but should be used sparingly at first. Some recommend butcher's offal; but, although ducks may be fattened on such food to an unusual weight, and thus be profitable for the market, their flesh will be rendered rank and gross, and not at all fit for the table.

To *fatten geese*, it is necessary to give them a little corn daily, with the addition of some raw Swedish turnips, carrots, mangel-wurtzel leaves, lucerne, tares, cabbage leaves, and lettuces. Barley-meal and water is recommended by some; but full-grown geese that have never been habituated to the mixture when young, will occasionally refuse to eat it. Cooked potatoes, in small quantities, do no harm; and, apart from the consideration of expense, steeped wheat would produce a firstrate delicacy.

Those who can only afford to bring up one or two, should confine them in a crib or some such place, about the beginning of July, and feed them as directed, giving them a daily supply of clean water for drink. If from a dozen to twenty are kept, a large pen of from fifteen to twenty feet square should be made, well covered with straw on the bottom, and a covered house in a corner for protection against the sun and rain, when required; since exposure to either of these is not good. It will be observed that, about noon, if geese are at liberty, they will seek some shady spot, to avoid the influence of the sun; and when confined in small places, they have not sufficient space for flapping their wings, and drying themselves after being wet, nor have they room for moving about so as to keep themselves warm. There should be three troughs in the crib: one for dry oats; another for vegetables, which ought always to be cut down; and a third for clean water, of which they must always have a plentiful supply. The riper the cabbages and lettuces are with which they are supplied the better.

Slaughtering and Dressing. Both ducks and geese should be led out to the pond a few hours before being slaughtered, where they will neatly purify and arrange their feathers. The common mode of slaughtering the latter—bleeding them from the internal parts of the throat—is needlessly slow and cruel.

Fowls for cooking, that are to be sent to a distance, or to be kept any time before being served, should be plucked, drawn, and dressed immediately after being killed. The feathers strip off much more easily and cleanly while the bird is yet warm. When large numbers are to be slaughtered and prepared in a short time, the process is expedited by scalding the bird in boiling water, when the feathers drop off almost at once. Fowls thus treated are, however, generally thought inferior in flavor, and are more likely to acquire a taint in close, warm weather, than such as are plucked and dressed dry.

In dressing, all bruises or rupturing of the skin should be avoided. A coarse, half-worn cloth, that is pervious to the air, like a wire sieve, and perfectly dry and clean, forms the best wrapper. The color of yellow-skinned turkeys—equally well-flavored, by the way—is improved for appearance at market by wrapping them for twelve or twenty-four hours in cloths soaked in cold salt and water, frequently changed. For the same purpose, the loose fat is first laid in warm salt and water, and afterward in milk and water for two or three hours. Some dust with flour, inside and out, any fowls that are to be carried far, or to hang many days before being cooked.

The oldest and toughest fowls, which are often pronounced unfit for eating, thrown away, and wasted, may be made into a savory and nutritious dish by jointing, after the bird is plucked and drawn, as for a pie; it should not be skinned. Stew it five hours in a close saucepan, with salt, mace, onions, or any other flavoring ingredients desired. When tender, turn it out into a deep dish, so that the meat may be entirely covered with the liquor. Let it stand thus in its own jelly for a day or two; it may then be served in the shape of a curry, a hash, or a pie, and will be found to furnish an agreeable repast.

Old geese, killed in the autumn, after they have recovered from moulting, and before they have begun to think about the breeding time, make excellent meat, if cut into small portions, stewed slowly five or six hours with savory condiments, and made into pie the next day. By roasting and broiling, the large quantity of nutriment contained in the bones and cartilages is lost, and what might easily be made tender has to be swallowed tough. Young geese, as well as the old, are, also, often salted and boiled.

POULTRY-HOUSES.

The three grand requisites in a poultry house are *cleanliness*, *dryness*, and *warmth*. A simple arrangement for this purpose is a shed built against the gable of the house, opposite to the part warmed by the kitchen fire, in which are placed cross-bars for roosting, with boxes for laying in, or quantities of fresh straw. This should always have an opening, to allow the poultry-house to be cleansed out, at least once a week. Fowls will never thrive long amidst uncleanliness; and even with the utmost care a place where they have been long kept becomes tainted, as it is called; the surface of the ground becomes saturated with their *exuriæ*, and is therefore no longer conducive to health.

To avoid this effect, some persons in the country frequently change the sites of their poultry-houses, to obtain fresh ground; while others, who cannot thus change, purify the houses by fumigations of blazing pitch, by washing with hot lime water, and by strewing large quantities of pure sand both within and without. Washing the floor every week is a necessity; for which purpose it is advantageous to have the house paved either with stones, bricks, or tiles. A good flooring, however, and cheaper than either of these, may be formed by using a composition of lime and smithy ashes, together with the riddlings of common kitchen ashes; these, having been all finely broken, must be mixed together with water, put on the floor with a mason's trowel, and nicely smoothed on the surface. If this is put on a floor which is in a tolerably dry situation, and allowed to harden before being used, it will become nearly as solid and compact as stone, and is almost as durable.

RUSTIC POULTRY-HOUSE.

The inside of the laying-boxes should be frequently washed with hot lime water, to free them from vermin, which greatly torment the sitting hens. For the same purpose, poultry should always have a heap of dry sand, or fine ashes, laid under some covered place or thick tree near their yard, in which they may dust themselves; this being their means of ridding themselves of the vermin with which they are annoyed.

In every esta l . men. or poultry-rearing, there ought to

be some separate crib or cribs, into which to remove fowl when laboring under disease; for, not only are many of the diseases to which poultry are liable highly contagious, but the sick birds are also regarded with dislike by such as are in health; and the latter will, generally, attack and maltreat them, aggravating, at least, their sufferings, if not actually depriving them of life. The moment, therefore, that a bird is perceived to droop, or appears pining, it should be removed to one of these infirmaries.

Separate pens are also necessary, to avoid quarrelling among some of the highly-blooded birds, more particularly the game fowl. They are also necessary when different varieties are kept, in order to avoid improper or undesirable commixture from accidental crossing. These lodgings may be most readily constructed in rows, parallel to each other; the partitions may be formed of lattice-work, being thus rather ornamental, and the cost of erection but trifling. Each of these lodgings should be divided into two compartments, one somewhat larger than the other; one to be close and warm, for the sleeping-room; and the other, a large one, airy and open, that the birds may enjoy themselves in the daytime. Both must be kept particularly dry and clean, and be well protected from the weather.

A FANCY COOP IN CHINESE OR GOTHIC STYLE.

A *hen-ladder* is an indispensable piece of furniture, though frequently absent. This is a sort of ascending scale of perches, one a little higher than the other; not exactly above its predecessor, but somewhat in advance. By neglecting the use of this very simple contrivance, many valuable fowls may be lost or severely injured, by attempting to fly down from their roost—an attempt from succeeding in which the birds are incapacitated, in consequence of the bulk of their body preponderating over the power of their wings.

Some people allow their fowl to roost abroad all night, in all weathers, in trees, or upon fences near the poultry-house. This is a slovenly mode of keeping even the humblest live stock; it offers a temptation to thieves, and the health of the fowls cannot be improved by their being soaked all night long in drenching rain, or having their feet frozen to the branches or rails. There is no difficulty in accustoming any sort of poultry, except the pea fowl, to regular housing at night.

It is better that turkeys should not roost in the same house with the domestic fowl, as they are apt to be cross to sitting and laying hens.

No poultry-house is what it ought to be, it may be suggested, in conclusion, unless it is in such a state as to afford a lady, without offending her sense of decent propriety, a respectable shelter on a showery day.

In our climate, the disorders to which poultry are liable are, comparatively, few in number, and they usually yield to judicious treatment. The little attention that has too generally been bestowed upon this subject may be accounted for from the circumstance that, in an economical point of view, the value of an individual fowl is relatively insignificant; and while the ailments of other domesticated animals generally claim a prompt and efficient care, the unhappy inhabitants of the poultry-yard are too often relieved of their sufferings in the most summary manner. There are reasons, however, which will justify a more careful regard in this matter, besides the humanity of adding to the comfort of these useful creatures; and the attempt to cure, in cases of disease, will often be rewarded by their flesh being rendered more palatable, and their eggs more wholesome.

Most of the diseases to which fowls are subject are the result of errors in diet or management, and should have been prevented, or may be removed by a change, and the adoption of a suitable regimen. When an individual is attacked, it should be forthwith removed, to prevent the contamination of the rest of the flock. Nature, who proves a guardian to fowls in health, will nurse them in their weakness, and act as a most efficient physician to the sick; and the aim of all medical treatment should be to follow the indications which Nature holds out, and assist in the effort which she constantly makes for the restoration of health.

The more common diseases which afflict poultry will be so described that they need not be misapprehended, and such remedies suggested as experience has proved to be salutary; and, taken alphabetically, the first on the list is

ASTHMA.

This common disease seems to differ sufficiently in its characteristics to warrant a distinction into two species. In one it appears to be caused by an obstruction of the air-cells, by an accumulation of phlegm, which interferes with the exercise of their functions. The fowl labors for breath, in consequence of not being able to take in the usual quantity of air at an inspiration. The capacity of the lungs is thereby diminished, the lining membrane of the windpipe becomes thickened, and its minute branches are more or less affected. These effects may, perhaps, be attributed to the fact that, as our poultry are originally natives of tropical climates, they require a more equal temperature than is afforded, except by artificial means, however well they may appear acclimated.

Another variety of asthma is induced by fright, or undue excitement. It is sometimes produced by chasing fowls to catch them, by seizing them suddenly, or by their fighting with each other. In these cases, a blood-vessel is often ruptured, and sometimes one or more of the air-cells. The *symptoms* are, short breathing; opening of the beak often, and for quite a time; heaving and panting of the chest; and, in case of a rupture of a blood-vessel, a drop of blood appearing on the beak.

Treatment. Confirmed asthma is difficult to cure. For the disease in its incipient state, the fowl should be kept warm, and treated with repeated doses of hippo-powder and sulphur, mixed with butter, with the addition of a small quantity of Cayenne pepper.

COSTIVENESS.

The existence of this disorder will become apparent by observing the unsuccessful attempts of the fowl to relieve itself. It frequently results from continued feeding on dry diet, without access to green vegetables. Indeed, without the use of these, or some substitute—such as mashed potatoes—costiveness is certain to ensue. The want of a sufficient supply of good water will also occasion the disease, on account of that peculiar structure of the fowl, which renders them unable to void their urine, except in connection with the *fæces* of solid food, and through the same channel.

Treatment. Soaked bread, with warm skimmed-milk, is a mild remedial agent, and will usually suffice. Boiled carrots or cabbage are more efficient. A meal of earth-worms is sometimes advisable; and hot potatoes, mixed with bacon-fat, are said to be excellent. Castor-oil and burned butter will remove the most obstinate cases; though a clyster of oil, in addition, may sometimes be required, in order to effect a cure.

DIARRHŒA.

There are times when fowls dung more loosely than at others, especially when they have been fed on green or soft food; but this may occur without the presence of disease. Should this state, however, deteriorate into a confirmed and continued laxity, immediate attention is required to guard against fatal effects. The *causes* of diarrhœa are dampness, undue acidity in the bowels, or the presence of irritating matter there.

The *symptoms* are lassitude and emaciation; and, in very severe cases, the voiding of calcareous matter, white, streaked with yellow. This resembles the yolk of a stale egg, and clings to the feathers near the vent. It becomes acrid, from the presence of ammonia, and causes inflammation, which speedily extends throughout the intestines.

Treatment. This, of course, depends upon the cause. If the disease is brought on by a diet of green or soft food, the food must be changed, and water sparingly given; if it arises from undue acidity, chalk mixed with meal is advantageous, but rice-flour boluses are most reliable. Alum-water, of moderate strength, is also beneficial. In cases of *bloody flux*, boiled rice and milk, given warm, with a little magnesia, or chalk, may be successfully used.

FEVER.

The most decided species of fever to which fowls are subject occurs at the period of hatching, when the animal heat is often so increased as to be perceptible to the touch. A state of fever may also be observed when they are about to lay. This is, generally, of small consequence, when the birds are otherwise healthy; but it is of moment, if any other disorder is present, since, in such case, the original malady will be aggravated. Fighting also frequently occasions fever, which sometimes proves fatal.

The *symptoms* are an increased circulation of the blood; excessive heat; and restlessness.

Treatment. Light food and change of air; and, if necessary, aperient medicine, such as castor oil, with a little burned butter.

INDIGESTION.

Cases of indigestion among fowls are common, and deserve attention according to the causes from which they proceed. A change of food will often produce *crop-sickness*, as it is called, when the fowl takes but little food, and suddenly loses flesh. Such disease is of little consequence, and shortly disappears. When it requires attention at all, all the symptoms will be removed by giving their diet in a warm state.

Sometimes, however, a fit of indigestion threatens severe consequences, especially if long continued. Every effort should be made to ascertain the cause, and the remedy must be governed by the circumstances of the case.

The *symptoms* are heaviness, moping, keeping away from the nest, and want of appetite.

Treatment. Lessen the quantity of food, and oblige the fowl to exercise in an open walk. Give some powdered cayenne and gentian, mixed with the usual food. Iron-rust, mixed with soft food, or diffused in water, is an excellent tonic, and is indicated when there is atrophy, or diminution of the flesh. It may be combined with oats or grain. Milk-warm ale has also a good effect, when added to the diet of diseased fowls.

PRAIRIE HENS.

LICE.

The whole feathered tribe seem to be peculiarly liable to be infested with lice; and there have been instances when fowls have been so covered in this loathsome manner that the natural color of the feathers has been undistinguishable. The presence of vermin is not only annoying to poultry, but materially interferes with their growth, and prevents their fattening. They are, indeed, the greatest drawback to the success and pleasure of the poultry fanciers; and nothing but unremitting vigilance will exterminate them, and keep them exterminated.

Treatment. To attain this, whitewash frequently all the parts adjacent to the roosting-pole, take the poles down and run them slowly through a fire made of wood shavings, dry weeds, or other light waste combustibles. Flour of sulphur, placed in a vessel, and set on fire in a close poultry-house, will penetrate every crevice, and effectually exterminate the vermin. When a hen comes off with her brood, the old nest should be cleaned out, and a new one placed; and dry tobacco-leaves, rubbed to a powder between the hands, and mixed with the hay of the nest, will add much to the health of the poultry.

Flour of sulphur may also be mixed with Indian-meal and water, and fed in the proportion of one pound of sulphur to two dozen fowls, in two parcels, two days apart. Almost any kind of grease, or unctuous matter, is also certain death to the vermin of domestic poultry. In the case of very young chickens, it should only be used in a warm, sunny day, when they should be put into a coop with their mother, the coop darkened for an hour or two, and every thing made quiet, that they may secure a good rest and nap after the fatigue occasioned by greasing them. They should be handled with great care, and greased thoroughly; the hen, also. After resting, they may be permitted to come out and bask in the sun; and in a few days they will look sprightly enough.

To guard against vermin, however, it should not be forgotten that *cleanliness* is of vital importance; and there must always be plenty of slacked lime, dry ashes, and sand, easy of access to the fowls, in which they can roll and dust themselves.

LOSS OF FEATHERS.

This disease, common to confined fowls, should not be confounded with the natural process of moulting. In this diseased state, no new feathers come to replace the old, but the fowl is left bald and naked; a sort of roughness also appears on the skin; there is a falling off in appetite, as well as moping and inactivity.

Treatment. As this affection is, in all probability, constitutional rather than local, external remedies may not always prove sufficient. Stimulants, however, applied externally, will serve to assist the operation of whatever medicine may be given. Sulphur may be thus applied, mixed with lard. Sulphur and cayenne, in the proportion of one quarter each, mixed with fresh butter, is good to be given internally, and will act as a powerful alterative. The diet should be changed; and cleanliness and fresh air are indispensable.

In *diseased moulting*, where the feathers stare and fall off, till the naked skin appears, sugar should be added to the water which the fowls drink, and corn and hemp-seed be given. They should be kept warm, and occasionally be treated to doses of cayenne pepper.

PIP.

This disorder, known also as the *gapes*, is the most common ailment of poultry and all domestic birds. It is especially the disease of young fowls, and is most prevalent in the hottest months, being not only troublesome but frequently fatal.

As to its *cause* and nature, there has been some diversity

of opinion. Some consider it a catarrhal inflammation, which produces a thickening of the membrane lining the nostrils and mouth, and particularly the tongue; others assert that it is caused by want of water, or by bad water; while others describe it as commencing in the form of a vesicle on the tip of the tongue, which occasions a thickened state of the skin, by the absorption of its contents. The better opinion, however, is, that the disease is occasioned by the presence of worms, or *fasciolæ*, in the windpipe. On the dissection of chickens dying with this disorder, the windpipe will be found to contain numerous small, red worms, about the size of a cambric needle, which, at the first glance, might be mistaken for blood-vessels. It is supposed by some that these worms continue to grow, until, by their enlargement, the windpipe is so filled up that the chicken is suffocated.

The common *symptoms* of this malady are the thickened state of the membrane of the tongue, particularly toward the tip; the breathing is impeded, and the beak is frequently held open, as if the creature were gasping for breath; the beak becomes yellow at its base; and the feathers on the head appear ruffled and disordered; the tongue is very dry; the appetite is not always impaired; but yet the fowl cannot eat, probably on account of the difficulty which the act involves, and sits in a corner, pining in solitude.

Treatment. Most recommend the immediate removal of the thickened membrane, which can be effected by anointing the part with butter or fresh cream. If necessary, the scab may be pricked with a needle. It will also be found beneficial to use a pill, composed of equal parts of scraped garlic and horse-radish, with as much cayenne pepper as will outweigh a grain of wheat; to be mixed with fresh butter, and given every morning; the fowl to be kept warm.

If the disease is in an advanced state, shown by the chicken's holding up its head and gaping for want of breath, the fowl should be thrown on its back, and while the neck is held straight, the bill should be opened, and a quill inserted into the windpipe, with a little turpentine. This being round, will loosen and destroy a number of small, red worms, some of which will be drawn up by the feather, and others will be coughed up by the chicken. The operation should be repeated the following day, if the gaping continues. If it ceases, the cure is effected.

It is stated, also, that the disease has been entirely prevented by mixing a small quantity of spirits of turpentine with the food of fowls; from five to ten drops, to a pint of meal, to be made into a dough. Another specific recommended is to keep iron standing in vinegar, and put a little of the liquid in the food every few days.

Some assert that it is promoted by simply scanting fowls in their food; and this upon the ground that chickens which are not confined with the hen, but both suffered to run at large and collect their own food, are not troubled with this disease. There can be little doubt that it is caused by inattention to cleanliness in the habits and lodgings of fowls; and some, therefore, think that if the chicken-houses and coops are kept clean, and frequently washed with thin whitewash, having plenty of salt and brine mixed with it, that it would be eradicated.

ROUP.

This disease is caused mainly by cold and moisture; but it is often ascribed to improper feeding and want of cleanliness and exercise. It affects fowls of all ages, and is either acute or chronic; sometimes commencing suddenly, on exposure; at others gradually, as the consequence of neglected colds, or damp weather or lodging. Chronic roup has been known to extend through two years. The most prominent *symptoms* are difficult and noisy breathing and gaping, terminating in a rattling in the throat; the head swells, and is feverish; the eyes are swollen, and the eye-lids appear livid; the sight decays, and sometimes total blindness ensues; there are discharges from the nostrils and mouth, at first thin and limpid, afterward thick, purulent, and fetid. In this stage, which resembles the glanders in horses, the disease becomes infectious.

SWANS.

As *secondary* symptoms, it may be noticed that the appetite fails, except for drink; the crop feels hard; the feathers are staring, ruffled, and without the gloss that appears in health; the fowl mopes by itself and seems to suffer much pain.

Treatment. The fowls should be kept warm, and have plenty of water and scalded bran, or other light food. When chronic, change of food and air is advisable. The ordinary remedies—such as salt dissolved in water—are inefficacious. A solution of sulphate of zinc, as an eye-water, is a valuable cleansing application. Rue-pills, and a decoction of rue, as a tonic, have been administered with apparent benefit.

The following is recommended: of powdered gentian and Jamaica ginger, each one part; Epsom salts, one and a half parts; and flour of sulphur, one part; to be made up with butter, and given every morning

The following method of treatment is practised by some of the most successful poulterers in the country: As soon as discovered, if in warm weather, remove the infected fowls to some well-ventilated apartment, or yard; if in winter, to some warm place; then give a dessert-spoonful of castor-oil; wash their heads with warm Castile-soap suds, and let them remain till next morning fasting. Scald for them Indian-meal, adding two and a half ounces of Epsom salts for ten hens, or in proportion for a less or larger number; give it warm, and repeat the dose in a day or two, if they do not recover.

Perhaps, however, the best mode of dealing with roup and all putrid affections is as follows: Take of finely pulverized, fresh-burnt charcoal, and of new yeast, each three parts; of pulverized sulphur, two parts; of flour, one part; of water, a sufficient quantity; mix well, and make into two doses, of the size of a hazel-nut, and give one three times a day. *Cleanliness* is no less necessary than warmth; and it will sometimes be desirable to bathe the eyes and nostrils with warm milk and water, or suds, as convenient.

WOUNDS AND SORES.

Fowls are exposed to wounds from many sources. In their frequent encounters with each other, they often result; the poultry-house is besieged by enemies at night, and, in spite of all precaution, rats, weasels, and other animals will assault the occupants of the roost, or nest, to their damage. These wounds, if neglected, often degenerate into painful and dangerous ulcers.

When such injuries occur, *cleanliness* is the first step toward a cure. The wound should be cleared from all foreign matter, washed with tepid milk and water, and excluded as far as possible from the air. The fowl should be removed from its companions, which, in such cases, seldom or never show any sympathy, but, on the contrary, are always ready to assault the invalid, and aggravate the injury. Should the wound not readily heal, but ulcerate, it may be bathed with alum-water. The ointment of creosote is said to be effectual, even when the ulcer exhibits a fungous character, or *proud flesh* is present. Ulcers may also be kept clean, if dressed with a little lard, or washed with a weak solution of sugar of lead; if they are indolent, they may be touched with blue-stone.

When severe *fractures* occur to the limbs of fowls, the best course, undoubtedly, to pursue—unless they are very valuable—is to kill them at once, as an act of humanity. When, however, it is deemed worth while to preserve them, splints may be used, when practicable. Great cleanliness must be observed; the diet should be reduced; and every precaution taken against the inflammation, which is sure to supervene. When it is established, cooling lotions—such as warm milk and water—may be applied.

Memoirs of the Queens of France. Written in France, carefully compiled from researches made there, commended by the press generally, and published from the Tenth London Edition. It is a truly valuable work for the reader and student of history. By Mrs. Forbes Bush. 2 vols. in one. Cloth. Price $1 75.

Memoirs of the Life of Anne Boleyn, Queen of Henry VIII. In the records of biography there is no character that more forcibly exemplifies the vanity of human ambition, or more thoroughly enlists the attention of the reader than this—the Seventh American, and from the Third London Edition. By Miss Benger. With portrait on steel. Cloth. $1 75.

Heroic Women of History. Containing the most extraordinary examples of female courage of ancient and modern times, and set before the wives, sisters, and daughters of the country, in the hope that it may make them even more renowned for resolution, fortitude, and self-sacrifice than the Spartan females of old. By Henry C. Watson. With Illustrations. Cloth. $1 75.

Public and Private History of Louis Napoleon, Emperor of the French. An impartial view of the public and private career of this extraordinary man, giving full information in regard to his most distinguished ministers, generals, relatives and favorites. By Samuel M. Schmucker, LL. D. With portraits on Steel. Cloth. $1 75.

Life and Reign of Nicholas I., Emperor of Russia. The only complete history of this great personage that has appeared in the English language, and furnishes interesting facts in connection with Russian society and government of great practical value to the attentive reader. By Samuel M. Schmucker, LL. D. With Illustrations. Cloth. $1 75.

Life and Times of George Washington. A concise and condensed narrative of Washington's career, especially adapted to the popular reader, and presented as the best matter upon this immortal theme—one especially worthy the attention and admiration of every American. By Samuel M. Schmucker, LL. D. With Portrait on steel. Cloth. $1 75.

LIFE AND TIMES OF ALEXANDER HAMILTON. Incidents of a career that will never lose its singular power to attract and instruct, while giving impressive lessons of the brightest elements of character, surrounded and assailed by the basest. By SAMUEL M. SCHMUCKER, LL. D. With Portrait on steel. Cloth. $1 75.

LIFE AND TIMES OF THOMAS JEFFERSON. In which the author has presented both the merits and defects of this great representative hero in their true light, and has studiously avoided indiscriminate praise or wholesale censure. By SAMUEL M. SCHMUCKER, LL. D. With Portrait. Cloth. $1 75.

LIFE OF BENJAMIN FRANKLIN. Furnishing a superior and comprehensive record of this celebrated Statesman and Philosopher—rich beyond parallel in lessons of wisdom for every age, calling and condition in life, public and private. By O. L. HOLLEY. With Portrait on steel and Illustrations on wood. Cloth. $1 75.

PUBLIC AND PRIVATE LIFE OF DANIEL WEBSTER. The most copious and attractive collection of personal memorials concerning the great Statesman that has hitherto been published, and by one whose intimate and confidential relations with him afford a guarantee for their authenticity. By Gen. S. P. LYMAN. With Illustrations. Cloth. $1 75.

LIFE AND TIMES OF HENRY CLAY. An impartial biography, presenting, by bold and simple strokes of the historic pencil, a portraiture of the illustrious theme which no one should fail to read, and no library be without. By SAMUEL M. SCHMUCKER, LL. D. With Portrait on steel. Cloth. $1 75.

LIFE AND PUBLIC SERVICES OF STEPHEN A. DOUGLAS. A true and faithful exposition of the leading incidents of his brilliant career arranged so as to instruct the reader and produce the careful study which the life of so great a man deserves. By H. M. FLINT. With Portrait on steel. Cloth. $1 75.

LIFE AND PUBLIC SERVICES OF ABRAHAM LIN-COLN. (In both the English and German languages.) As a record of this great man it is a most desirable work, admirably arranged for reference, with an index over each page, from which the reader can familiarize himself with the contents by glancing through it. By FRANK CROSBY, of the Philadelphia Bar. With Portrait on steel. Cloth. $1 75.

LIFE OF DANIEL BOONE, THE GREAT WESTERN HUNTER AND PIONEER. Comprising graphic and authentic accounts of his daring, thrilling adventures, wonderful skill, coolness and sagacity under the most hazardous circumstances, with an autobiography dictated by himself. By CECIL B. HARTLEY. With Illustrations. Cloth. $1 75.

LIFE OF COLONEL DAVID CROCKET, THE ORIGINAL HUMORIST AND IRREPRESSIBLE BACKWOODSMAN. Showing his strong will and indomitable spirit, his bear hunting, his military services, his career in Congress, and his triumphal tour through the States—written by himself; to which is added the account of his glorious death at the Alamo. With Illustrations. Cloth. $1 75.

LIFE OF KIT CARSON, THE GREAT WESTERN HUNTER AND GUIDE. An exciting volume of wild and romantic exploits, thrilling adventures, hair-breadth escapes, daring coolness, moral and physical courage, and invaluable services—such as rarely transpire in the history of the world. By CHARLES BURDETT. With Illustrations. Cloth. $1 75.

LIFE OF CAPTAIN JOHN SMITH, THE FOUNDER OF VIRGINIA. The adventures contained herein serve to denote the more noble and daring events of a period distinguished by its spirit, its courage, and its passion, and challenges the attention of the American people. By W. GILMORE SIMMS. With Illustrations. Price $1 75.

LIFE OF GENERAL FRANCIS MARION, THE CELEBRATED PARTISAN HERO OF THE REVOLUTION. This was one of the most distinguished men who figured on the grand theatre of war during the times that "tried men's souls," and his brilliant career has scarcely a parallel in history. By CECIL B. HARTLEY. With Illustrations. Cloth. $1 75.